ATLAS OF HUMAN PARASITOLOGY

ATLAS OF HUMAN PARASITOLOGY

Lawrence R. Ash, PhD

Professor of Infectious and Tropical Diseases
Associate Dean and Chairman
School of Public Health
University of California, Los Angeles
Los Angeles, California

Thomas C. Orihel, PhD

Professor of Parasitology
Department of Tropical Medicine
School of Public Health and Tropical Medicine
Tulane University
New Orleans, Louisiana

Educational Products Division
American Society of Clinical Pathologists
Chicago

Cover Plate

Top row. **Left,** *Clonorchis sinensis,* adult Chinese liver fluke, carmine stain. **Center,** *Trichuris vulpis,* egg in feces, unstained. **Right,** *Ascaris lumbricoides,* egg in feces, unstained.

Middle row. **Center,** *Diphyllobothrium latum,* egg in feces, unstained. **Right,** *Hymenolepis nana,* egg in feces, unstained.

Bottom row. **Left,** *Wuchereria bancrofti,* microfilaria in thick blood smear, Giemsa's stain. **Center,** *Trichomonas vaginalis,* trophozoite from culture, Giemsa's stain. **Right,** *Entamoeba coli,* trophozoite in feces, trichrome stain.

Library of Congress Cataloging in Publication Data

Ash, Lawrence R.   1933–
    Atlas of human parasitology.

    Bibliography: p.
    Includes index.
    1. Medical parasitology—Atlases. I. Orihel,
Thomas C., 1929–   joint author.   II. Title.
[DNLM: 1. Parasites—Atlases. 2. Parasitic
diseases—Atlases. QX 17 A819a]
RC119.A8   616.9'6   80-25291
ISBN O-89189-081-5

Table of Contents

vi   Table of Contents

List of Plates

Preface

The major objective of this atlas is to provide students, laboratory personnel, and the physician-pathologist with a guide to the identification of human parasites. An effort has been made to include all parasite groups, save one, that naturally occur in humans, as well as some that are uncommon or rare. The exception is the malaria parasite, which we have opted to exclude for a variety of reasons. Because of technical limitations, photographs might not capture the great range of morphologic variation in a given stage of the malaria parasite, the important color variations, and the significant—but subtle—morphologic features. Hence, we are convinced that the fine, hand-drawn, color illustrations currently available and widely reproduced in medical and parasitologic references remain the best source of information for identification of these parasites.

It has been our teaching experience that narrative descriptions of morphologic features of parasites often are difficult for the inexperienced person to comprehend accurately without the aid of good illustrations, either photographs or drawings. In a sense, reading is not the same as seeing. Cysts, trophozoites, eggs, microfilariae, and other parasitic forms, properly concentrated or otherwise prepared for microscopic study, typically are unique in appearance, and their identification usually is straightforward. Once the reader sees them and studies their characteristics, he or she is not likely to forget them. This is not to say that, with certain related organisms, accurate identifications do not pose problems. Subtle variations in morphology are more easily demonstrated with visual material than through written descriptions. Color often is a useful and important criterion and, in these circumstances, the color photograph is a desirable vehicle for conveying such information.

This atlas is not intended to be a substitute for standard reference textbooks. The text materials presented here are abbreviated and only highlight the important facts regarding the parasites. For additional, detailed information, the reader is obliged to consult the standard references. Specific details on the performance of diagnostic procedures can be obtained from some of the standard textbooks or from manuals devoted exclusively to diagnostic parasitology.

This book is intended for use primarily in the laboratory. The parasite stages are presented as they would normally be seen in standard diagnostic and clinical laboratory procedures. In virtually all instances, the photographs were prepared from clinical materials processed by standard procedures.

Acknowledgments

This atlas has evolved, with numerous changes and the addition of new and expanded material, from an earlier effort in which the authors collaborated with four colleagues, Drs James W. Smith, Russell M. McQuay, Dorothy Mae Melvin, and John H. Thompson, Jr, in producing *Diagnostic Medical Parasitology,* a three-volume series that provided kodachrome slides and text material on the diagnostic stages of parasites affecting humans. We are indebted to these colleagues for their encouragement, suggestions, and the contribution of added materials that have been of great help in the production of this atlas.

We also wish to acknowledge and thank our many other colleagues who provided material for photomicrography or, in some cases, kodachrome slides from which some of our illustrations were prepared. These include, in alphabetical order, Marilyn Bartlett, Paul C. Beaver, Daniel H. Connor, Mark Eberhard, Lynne S. Garcia, A. O. Heydorn, Maurice D. Little, Emile A. Malek, Michel Rommel, John F. Schacher, Marietta Voge, Edward Wagner, and Robert G. Yaeger.

Special credit and thanks are due Jeffery L. Smith, Department of Tropical Medicine, Tulane University School of Public Health and Tropical Medicine, who prepared the prints for the atlas. His technical skills, attention to detail, and standard of excellence, as well as his many hours spent in the darkroom processing prints for the authors, have contributed immeasurably to the production of this book. We also are indebted to Max Lent, Division of Epidemiology, UCLA School of Public Health, for his invaluable technical assistance in the photomicrographic aspects of this atlas. We would like to thank Barbara van Duym, Parasitology Laboratory, Tulane University, Delta Regional Primate Research Center, for her assistance in preparing many of the line drawings used in the text.

We are grateful to Velia Irvine and Virginia Hansen for their special skills and diligence in preparation of the manuscript.

Finally, we would like to acknowledge and thank Dr William Mahoney, Patricia Lorimer, Patricia Klegman, Karen Izui, and the rest of the editorial staff of the Educational Products Division of the American Society of Clinical Pathologists for their considerable encouragement, enthusiasm, help, patience, and perseverance in bringing this atlas to fruition.

LAWRENCE R. ASH
THOMAS C. ORIHEL

Introduction

Parasitic diseases are widely distributed around the world, and the morbidity they produce in human populations has an important impact on the economic development of many countries. Many new international programs have been organized to combat these diseases, which often are referred to as the neglected diseases of humans. In many of the more developed countries, such as the United States, parasitic diseases are no longer endemic problems. However, with improved, world-wide transportation available to large segments of the world's population, parasitic diseases are carried across borders freely by travelers, temporary residents, and immigrants. Parasites once regarded as "exotic" no longer are considered so. Although most parasitic infections are not likely to become established in developed countries where high levels of sanitation are maintained, infected persons frequently require and seek medical treatment for them.

Often the parasites simply are incidental findings made in the course of routine laboratory work-ups. Technical personnel in the clinical laboratory must be prepared to recognize all parasites that are common in humans, as well as those considered more "exotic." As parasitic infections in humans assume a decreasing importance in our own country, less material is available for study and less time is devoted to training. Yet, laboratory personnel must maintain a high level of proficiency in parasite identification. Maintenance of this proficiency can be frustrating when parasite materials are so scarce in the laboratory. Consequently, although there are many parasites rarely observed by clinical laboratory personnel, workers must be prepared to identify them when circumstances require it.

The parasite stages presented here are typical and have not been idealized in any way. For the most part, they have been obtained from specimens submitted for routine examination. The majority of the intestinal protozoa depicted in this atlas have been photographed in fecal smears stained with trichrome or iron hematoxylin, although some organisms are illustrated as they appear in temporary wet mount preparations stained with iodine or unstained. Giemsa's stain has been the stain of choice for the blood and tissue protozoa, whereas both Giemsa's and hematoxylin stains have been used for microfilariae in blood smears. The photographs of helminth eggs and larvae are as they appear in saline or iodine-stained temporary wet mounts.

Sizes of parasitic elements are not provided with each illustration. The measurements, including ranges and means, are found in the text description of each parasite, as well as in summary tables. In general, the protozoan parasites, with the exception of *Balantidium*, were photographed under oil immersion, whereas helminth eggs and microfilariae were photographed at low (10×) and high (40×) magnification. In some instances, it will be apparent that elements were photographed under higher magnification to show details of structural features. An effort has been made to maintain consistency in size relationships within plates and among similar or related parasites. It will be obvious when this is not the case, and in most instances, it will be pointed out in the legends as well.

With few exceptions, the illustrations are of human material. Animal material has been used only rarely—generally when the parasite is exceedingly difficult to obtain from humans, eg, trypanosomes. The diagnostic stages of animal parasites that also are important to humans have been included, even though these stages do not occur in humans.

The taxonomy and nomenclature used in the atlas are based on currently accepted usage. We have not attempted to resolve areas of controversy.

Part I

Procedures

Parasite Diagnostic Procedures

A number of texts and laboratory manuals cover in detail the diagnosis of parasitic infections (see "Recommended References"). However, since the identification of parasitic organisms in feces, urine, blood, and tissues is dependent upon the application of proper diagnostic methods, a brief description of the most commonly used procedures is given.

Fecal Examination

SPECIMEN COLLECTION AND HANDLING.—Feces should be collected into clean, wide-mouthed containers and should not be contaminated with urine, water, or soil. Specimens collected following use of barium or oily laxatives generally are unsuitable for parasitologic examination. Liquid specimens should be examined within 30 minutes of passage, or the material should be placed in an appropriate preservative. Formed specimens may be kept at room temperature for several hours before examination, but should be examined the same day. If they must be kept longer, however, the specimens should be refrigerated (3–5 C) and portions should be put into preservative. Fecal specimens should never be placed into an incubator. Multiple specimens collected over a period of several days to a week may be necessary to optimize diagnosis of some parasitic infections (amebiasis, giardiasis, strongyloidiasis, and others).

SPECIMEN PROCESSING.—Several options are available in fecal specimen examination, depending upon the capabilities of the laboratory, the time available, and the information desired. Among the considerations are:

Description.—The specimen is described in terms of consistency (watery, loose, soft, or formed).

Wet mount examinations.—Saline wet mounts are most useful for the detection of motile trophozoites of amebae and flagellates in liquid specimens, but this material usually must be examined within 30 minutes of passage. Although protozoan cysts and helminth eggs and larvae can be detected in such preparations, successful detection generally is related directly to the intensity of the infection. Iodine wet mounts are especially useful for staining glycogen and making nuclei visible in protozoan cysts. Many laboratories have eliminated wet mount examinations, especially with formed stools, in favor of concentration procedures and/or permanently stained slides of fecal smears.

Concentration procedures.—The two most commonly used concentration procedures are the zinc sulfate flotation and the formalin-ether sedimentation methods. The former is used on fresh specimens (specific gravity of solution should be 1.18), or, more rarely, on formalin-preserved specimens (specific gravity of solution should be 1.20). Formalin-ether sedimentation is useful for both fresh and formalin-preserved material. This method also can be used on material preserved with polyvinyl alcohol (PVA), but with some decreased efficiency. Problems associated with the safety of keeping diethyl ether in laboratories have led to the development of alternative solvents for use in the formalin-ether concentration procedure. Ethyl acetate has been shown to be a safe and satisfactory substitute for diethyl ether. Material from either the flotation or sedimentation procedures is examined in temporary wet mount preparations.

Permanent stains.—Permanent-stained slides of fecal smears are used primarily for the diagnosis of intestinal protozoan infections. Smears made from fresh feces usually are fixed in Schaudinn's fixative prior to staining. PVA-fixed smears are made directly from feces preserved in PVA, and no further fixation is necessary prior to staining. The two most widely used permanent stains are trichrome and iron hematoxylin or various modifications of the hematoxylin stain.

With trichrome stain, the cytoplasm of organisms stains a blue-green tinged with purple, and nuclear chromatin, chromatoid material, erythrocytes, and bacteria stain red or purplish-red. Background material will stain green to blue-green. Organisms undergoing degeneration take a pale stain, and incompletely fixed organisms may stain red. In hematoxylin stains, the organisms stain blue-black, purplish, or bluish, depending on the exact method employed. Nuclear structures, chromatoid bodies, bacteria, and erythrocytes stain darkly. Background fecal material stains black or bluish-purple.

PRESERVATION OF FECES.—When fresh fecal specimens cannot be promptly or properly examined, all—or portions of—the specimen must be preserved in appropriate fixatives. Although many fixatives are available for use, only PVA and formalin will be mentioned here.

Polyvinyl alcohol solution is highly recommended for preservation of feces. PVA, a mixture of a fixative and a water-soluble resin, is the only preservative that allows for easy preparation of fecal smears for permanent staining. Its greatest use is for the demonstration of intestinal protozoan infections.

The use of 5% or 10% formalin or formol-saline is recommended as a preservative for fecal specimens, either for bulk fixation or fixation following concentration of fecal material. Although formalin-fixed material is well suited for use in the formalin-ether sedimentation technique, it is not useful for preparation of permanently stained smears.

SPECIAL PROCEDURES FOR FECAL MATERIAL. — A host of special procedures are available to aid in the diagnosis of specific protozoan and helminthic infections. Inoculation of culture media is occasionally helpful for some protozoa. Special procedures with helminths include cultures of larval stages, hatching techniques for eggs, and methods for the examination of large amounts of feces at one time (Kato thick smear technique).

Blood Examination

Although trypanosomes and filarial infections sometimes can be diagnosed by examination of fresh blood, the regular methods of blood examination involve the preparation of stained thick or thin blood films, or the use of blood concentration procedures. Microscopic diagnosis of malaria is by examination of stained thick or thin blood films.

PREPARATION OF THICK AND THIN BLOOD FILMS. — For thin blood films, a small drop of blood is spread over a slide in a thin layer, and following fixation and staining, all the blood components — erythrocytes, white cells, and platelets — are intact. Giemsa's stain, rather than Wright's stain, is recommended. At the terminal, feathered end of the smear, the cells are only one layer thick, allowing easy visualization of malaria-infected red blood cells, trypanosomes, or microfilariae. The process of making the thin film frequently results in organisms, such as trypanosomes and microfilariae, being carried to the end of the smear, or more often occurring at the edges of the film.

Thick blood films allow for examination of a much larger amount of blood, as the blood is concentrated into a smaller area, several cell layers thick. Prior to staining, the thick film must be lysed so that only white blood cells, platelets, ghosts of red blood cells, and parasites are visible following staining.

CONCENTRATION PROCEDURES. — The two blood concentration methods most widely used for demonstrating microfilariae are the Knott concentration technique and membrane filtration. Microfilariae isolated by the Knott concentration technique can be stained permanently, hence, only this method will be described.

The Knott concentration procedure, used to detect microfilariae in blood, consists of placing 1 ml of blood in 10 ml of 2% formalin. The formalin solution lyses the red blood cells, so that when the suspension is centrifuged, the sediment will contain only white blood cells and microfilariae, if present. These preparations can be examined as temporary wet mounts or the fluid can be allowed to dry on the slide and then stained with Giemsa's or hematoxylin stain.

Parasite Identification

In addition to utilizing good microscopic techniques, an essential and indispensable aid in diagnostic parasitology is the ability to do precise measurements using a calibrated ocular micrometer. Size of organisms frequently is a major consideration in differential diagnosis of parasite species. In this brief section, the characteristic structure of protozoa and helminths, and the useful features for their diagnoses are summarized.

Diagnosis of Protozoan Infections

Identification of the intestinal protozoan parasites in humans depends mainly on the recognition of their cyst and/or trophozoite stages. The latter stage characteristically has a thin, limiting membrane and exhibits considerable variation in size and shape. Cysts are spherical, subspherical, or somewhat elongated, showing less size variation and having a smooth, uniform wall.

Morphologic characteristics used to identify intestinal and atrial protozoa include:

Size. — Following fixation, some shrinkage of trophozoites and cysts usually occurs so that measurements of living organisms generally are somewhat greater than those in fixed material.

Motility. — Trophozoites of amebae, flagellates, and the ciliate *Balantidium coli* may exhibit characteristic movement in freshly passed liquid or soft specimens. Protozoan trophozoites rarely are found in formed feces. Some amebae, such as *Entamoeba histolytica*, may have a progressive, directional movement, whereas others may have a slower, random movement (eg, *E. coli* and *Endolimax nana*). The flagellates *Giardia*, *Chilomastix*, *Trichomonas*, and *Dientamoeba* all tend to have rather characteristic motility, as does the ciliate *B. coli*.

Nuclei. — Most trophozoites have a single nucleus, with the exception of the flagellates *Dientamoeba* and *Giardia*, and the ciliate *B. coli*. Depending on the species, cysts may have from one to eight nuclei, with mature cysts having a characteristic number. Nuclear structure is highly important in species diagnosis. Important considerations of nuclear structure include the size and location of the karyosome, the presence or absence of peripheral chromatin on the inner surface of the nuclear membrane, as well as its pattern of distribution, and the presence of additional chromatin material within the nucleus (Fig 1).

Cytoplasm. — The appearance of the cytoplasm, especially in trophozoites, may be useful in diagnosis. It may be coarsely or finely granular, may contain vacuoles, fibrils or organelles, and ingested material (erythrocytes, white blood cells, bacteria, yeasts). Cysts of amebae may contain chromatoid bodies and glycogen; flagellates may contain fibrils.

AMEBAE

The individual morphologic characteristics of the trophozoites and cysts of amebae are summarized in Tables 1 and 2, as well as in the individual parasite descriptions. The ability to identify amebae based on nuclear structure in saline or iodine wet mount preparations is limited. It occasionally is possible to recognize the nuclei of *Entamoeba coli* cysts and trophozoites in saline wet mounts. In cysts of *E. histolytica*, *E. coli*, *E. hartmanni*, and *Endolimax nana* it is possible to at least enumerate them in iodine wet mounts. Iodine is especially useful for staining glycogen and making nuclei visible in temporary preparations.

Since a specific diagnosis usually depends on the morphologic features of the nucleus, preparation of stained slides is essential, particularly when the presence of the pathogenic ameba, *E. histolytica*, is in question. Variations in nuclear structure not only are a normal biologic occurrence, but may be prompted by environmental conditions, as well as delays in fixation. In a differential diagnosis of *E. histolytica* and *E. coli*, the karyosome of *E. histolytica* is not always centrally located, and peripheral chromatin is not always finely and evenly distributed. Moreover, the karyosome in *E. coli* is not always eccentric, and peripheral chromatin is not consistently irregular in size and distribution.

If, in a stained smear, the preponderance of organisms is designated unequivocally as one or the other species, along with occasional organisms displaying atypical morphologic patterns, these organisms all may be of the same species, the rare forms simply having aberrant nuclear structure. On the other hand, one species may abound and another may be present in such low numbers that their de-

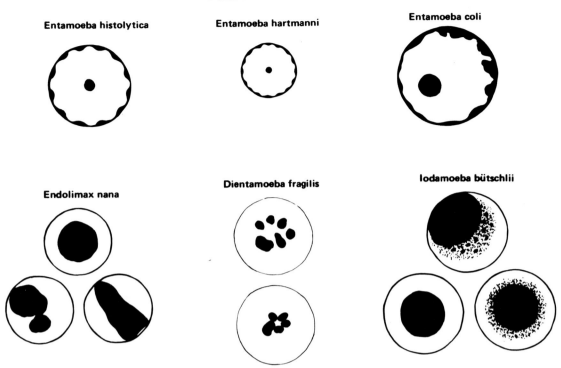

Fig 1. Nuclei of intestinal protozoa. From Smith et al, 1976b. Used by permission.

tection and proper recognition will be masked.

A cardinal rule is that a specific diagnosis of every organism found is not possible. However, the presence of a single species is indicative of the fact that the person has ingested fecally contaminated material, and—whether or not immediately recognized—additional organisms may be present.

Cysts of the amebae usually are finely granular. When mature, these may contain one *(Iodamoeba)*, four *(E. histolytica, E. hartmanni, E. nana)*, or eight nuclei *(E. coli)*. Glycogen masses, most common in cysts of *Iodamoeba*, also may occur in immature cysts of *Entamoeba* species and *E. nana*. Other cytoplasmic inclusions found in cysts include chromatoid bodies that characteristically are rod-shaped with rounded ends *(E. histolytica* and *E. hartmanni)* or with jagged ends *(E. coli)*. These do not stain with iodine.

Two nonpathogenic intestinal or atrial amebae are not illustrated in this atlas. *E. gingivalis*, which exists only in the trophozoite stage in the oral cavity, is rarely looked for or seen. *E. polecki*, which is the intestinal species, occurs mainly in pigs and is transmitted to humans living in close association with these animals. *E. polecki* tends to be a focal infection in many areas of the world. It has both trophozoite and cyst stages; the cyst always is uninucleate and usually contains many chromatoid bodies.

In recent years, several genera of free-living amebae have been well documented as important incidental human pathogens. The genera *Naegleria, Hartmannella,* and *Acanthamoeba* all have been implicated in human infections. *Naegleria* apparently produces a fulminating, usually fatal, primary amebic meningoencephalitis, whereas species of the other genera also may produce chronic infections, although these often are fatal. Trophozoites may be found in the cerebrospinal fluid in acute infections. These organisms are characterized by nuclei with very large karyosomes and no peripheral chromatin, closely resembling the nuclei seen in *Iodamoeba*.

Only trophozoites of *Naegleria* can be found in brain tissue, whereas both trophozoites and cysts of *Acanthamoeba* may occur. Tissue stages of these amebae also are characterized and best identified by finding nuclei with large karyosomes.

The amebae illustrated in this atlas include: *Entamoeba histolytica, E. hartmanni, E. coli, Endolimax nana, Iodamoeba bütschlii,* and *Naegleria fowleri.*

FLAGELLATES

INTESTINAL FLAGELLATES.—The sizes and morphologic characteristics of the intestinal and atrial flagellate parasites are summarized in Tables 3 and 4. *Giardia lamblia, Chilomastix mesnili,* and *Trichomonas* species are the best known flagellates.

TABLE 1.—Morphology of Trophozoites of Intestinal Amebae*

SPECIES	SIZE (DIAMETER OR LENGTH)	MOTILITY	NUCLEUS			CYTOPLASM	
			NUMBER	PERIPHERAL CHROMATIN	KARYOSOMAL CHROMATIN	APPEARANCE	INCLUSIONS
Entamoeba histolytica	10–60 μ. Usual range, 15–20 μ – commensal form †. Over 20 μ – invasive form.‡	Progressive, with hyaline, finger-like pseudopods.	1. Not visible in unstained preparations.	Fine granules. Usually evenly distributed and uniform in size.	Small, discrete. Usually centrally located, but occasionally is eccentric.	Finely granular.	Erythrocytes occasionally. Noninvasive organisms may contain bacteria.
Entamoeba hartmanni	5–12 μ. Usual range, 8–10 μ.	Usually nonprogressive, but may be progressive occasionally.	1. Not visible in unstained preparations.	Similar to *E. histolytica.*	Small, discrete, often eccentrically located.	Finely granular.	Bacteria.
Entamoeba coli	15–50 μ. Usual range, 20–25 μ.	Sluggish, nonprogressive, with blunt pseudopods.	1. Often visible in unstained preparations.	Coarse granules, irregular in size and distribution.	Large, discrete, usually eccentrically located.	Coarse, often vacuolated.	Bacteria, yeasts, other materials.
Entamoeba polecki	10–25 μ. Usual range, 15–20 μ.	Usually sluggish, similar to *E. coli.* Occasionally in diarrheic specimens, motility may be progressive.	1. May be slightly visible in unstained preparations. Occasionally distorted by pressure from vacuoles in cytoplasm.	Usually, fine granules evenly distributed. Occasionally, granules may be irregularly arranged. Chromatin sometimes in plaques or crescents.	Small, discrete, eccentrically located. Occasionally large, diffuse, or irregular.	Coarsely granular, may resemble *E. coli.* Contains numerous vacuoles.	Bacteria, yeasts.
Endolimax nana	6–12 μ. Usual range, 8–10 μ.	Sluggish, usually nonprogressive, with blunt pseudopods.	1. Visible occasionally in unstained preparations.	None.	Large, irregularly shaped, blotlike.	Granular, vacuolated.	Bacteria.
Iodamoeba bütschlii	8–20 μ. Usual range, 12–15 μ.	Sluggish, usually nonprogressive.	1. Not usually visible in unstained preparations.	None.	Large, usually centrally located. Surrounded by refractile, achromatic granules. These granules are often not distinct even in stained slides.	Coarsely granular, vacuolated.	Bacteria, yeasts, or other material.

*From Smith et al, 1976b. Adapted with permission from Brooke and Melvin: Morphology of Diagnostic Stages of Intestinal Parasites of Man, USDHEW PHS publication no. 1966, 1969.
†Usually found in asymptomatic or chronic cases; may contain bacteria.
‡Usually found in acute cases; often contain red blood cells.

TABLE 2. — Morphology of Cysts of Intestinal Amebae*

| SPECIES | SIZE | SHAPE | NUCLEUS | | | CHROMATOID BODIES | CYTOPLASM |
			NUMBER	PERIPHERAL CHROMATIN	KARYOSOMAL CHROMATIN		GLYCOGEN
Entamoeba histolytica	10–20 μ. Usual range, 12–15 μ.	Usually spherical.	4 in mature cyst. Immature cysts with 1 or 2 occasionally seen.	Peripheral chromatin present. Fine, uniform granules, evenly distributed.	Small, discrete, usually centrally located.	Present. Elongated bars with bluntly rounded ends.	Usually diffuse. Concentrated mass often present in young cysts. Stains reddish brown with iodine.
Entamoeba hartmanni	5–10 μ. Usual range, 6–8 μ.	Usually spherical.	4 in mature cyst. Immature cysts with 1 or 2 often seen.	Similar to *E. histolytica*.	Similar to *E. histolytica*.	Present. Elongated bars with bluntly rounded ends.	Similar to *E. histolytica*.
Entamoeba coli	10–35 μ. Usual range, 15–25 μ.	Usually spherical. Occasionally oval, triangular, or of another shape.	8 in mature cyst. Occasionally, supernucleate cysts with 16 or more are seen. Immature cysts with 2 or more occasionally seen.	Peripheral chromatin present. Coarse granules irregular in size and distribution, but often appear more uniform than in trophozoites.	Large, discrete, usually eccentrically, but occasionally centrally located.	Present, but less frequently seen than in *E. histolytica*. Usually splinterlike with pointed ends.	Usually diffuse, but occasionally well defined mass in immature cysts. Stains reddish brown with iodine.
Entamoeba polecki	9–18 μ. Usual range, 11–15 μ.	Spherical or oval.	1; rarely 2. Occasionally visible in unstained preparations.	Usually, fine granules evenly distributed.	Usually small and eccentrically located.	Present. Many small bodies with angular or pointed ends, or few large ones. May be oval, rodlike or irregular.	Usually small, diffuse masses. Stains reddish brown with iodine. A dark area called inclusion mass (possibly concentrated cytoplasm) is often also present. Mass does not stain with iodine.
Endolimax nana	5–10 μ. Usual range, 6–8 μ.	Spherical, ovoid, or ellipsoidal.	4 in mature cysts. Immature cysts with less than 4 rarely seen.	None.	Large (blot-like), usually centrally located.	Occasionally, granules or small oval masses seen, but bodies as seen in *Entamoeba* species are not present.	Usually diffuse. Concentrated mass seen occasionally in young cysts. Stains reddish brown with iodine.
Iodamoeba bütschlii	5–20 μ. Usual range, 10–12 μ.	Ovoid, ellipsoidal, triangular, or of another shape.	1 in mature cyst.	None.	Large, usually eccentrically located. Refractile, achromatic granules on one side of karyosome. Indistinct in iodine preparations.	Granules occasionally present, but chromatoid bodies as seen in *Entamoeba* species are not present.	Compact, well defined mass. Stains dark brown with iodine.

*From Smith et al, 1976b. Adapted with permission from Brooke and Melvin: Morphology of Diagnostic Stages of Intestinal Parasites of Man, USDHEW PHS publication no. 1966, 1969.

TABLE 3.—Morphology of Trophozoites of Intestinal Flagellates*

SPECIES	LENGTH	SHAPE	MOTILITY	NUMBER OF NUCLEI	NUMBER OF FLAGELLA†	OTHER FEATURES
Dientamoeba fragilis	5–15 μ. Usual range, 9–12 μ.	Ameboid. Pseudopodia are angular, serrated, or broad-lobed and hyaline, almost transparent.	Sluggish.	1 or 2. In approximately 40% of organisms only 1 nucleus is present. Nuclei not visible in unstained preparations.	None	Karysome usually in form of cluster of 4–8 granules. No peripheral chromatin. Cytoplasm is finely granular, vacuolated, and may contain bacteria. Organism formerly classified as an ameba.
Trichomonas hominis	8–20 μ. Usual range, 11–12 μ.	Pear-shaped.	Rapid, jerking.	1. Not visible in unstained mounts.	3–5 anterior; 1 posterior.	Undulating membrane extending length of body.
Trichomonas vaginalis	7–23 μ. Usual range, 11–15 μ.	Pear-shaped.	Rapid, jerking.	1. Not visible in unstained mounts.	3–5 anterior; 1 posterior.	Undulating membrane extends ½ length of body. No free posterior flagellum; does not live in intestinal tract; seen in vaginal smears and urethral discharges.
Chilomastix mesnili	6–24 μ. Usual range, 10–15 μ.	Pear-shaped.	Stiff, rotary.	1. Not visible in unstained mounts.	3 anterior; 1 in cytostome.	Prominent cytostome extending ⅓–½ length of body. Spiral groove across ventral surface.
Giardia lamblia	10–20 μ. Usual range, 12–15 μ.	Pear-shaped.	"Falling leaf."	2. Not visible in unstained mounts.	4 lateral; 2 ventral; 2 caudal.	Sucking disk occupying ½–¾ of ventral surface.
Enteromonas hominis	4–10 μ. Usual range, 8–9 μ.	Oval.	Jerking.	1. Not visible in unstained mounts.	3 anterior; 1 posterior.	One side of body flattened. Posterior flagellum extending free, posteriorly or laterally.
Retortamonas intestinalis	4–9 μ. Usual range, 6–7 μ.	Pear-shaped or oval.	Jerking.	1. Not visible in unstained mounts.	1 anterior; 1 posterior.	Prominent cytostome extending approximately ½ length of body.

*From Smith et al, 1976b. Adapted with permission from Brooke and Melvin: Morphology of Diagnostic Stages of Intestinal Parasites of Man, USDHEW PHS publication no. 1966, 1969.
†Not a practical feature for identification of species in routine fecal examination.

TABLE 4. — Morphology of Cysts of Intestinal Flagellates[°]

SPECIES	SIZE	SHAPE	NUMBER OF NUCLEI	OTHER FEATURES
Dientamoeba fragilis	No cyst.	—	—	—
Trichomonas hominis	No cyst.	—	—	—
Trichomonas vaginalis	No cyst.	—	—	—
Chilomastix mesnili	6–10 μ. Usual range, 7–9 μ.	Lemon-shaped, with anterior hyaline knob or "nipple."	1. Not visible in unstained preparations.	Cytostome with supporting fibrils. Usually visible in stained preparation.
Giardia lamblia	8–19 μ. Usual range, 11–12 μ.	Oval or ellipsoid.	Usually 4. Not distinct in unstained preparations. Usually located at one end.	Fibrils and flagella oriented longitudinally in cyst with other deep-staining fibrils lying laterally or obliquely in lower part of cyst. Cytoplasm often retracted from cyst wall. There may also be a "halo" effect around outside of cyst wall in stained smears.
Enteromonas hominis	4–10 μ. Usual range, 6–8 μ.	Elongate or oval.	1–4; usually 2 lying at opposite ends of cyst. Not visible in unstained mounts.	Resembles *E. nana* cyst. Fibrils or flagella are usually not seen.
Retortamonas intestinalis	4–9 μ. Usual range, 4–7 μ.	Pear-shaped or slightly lemon-shaped.	1. Not visible in unstained mounts.	Resembles *Chilomastix* cyst. Shadow outline of cytostome with supporting fibrils extends above nucleus.

[°]From Smith et al, 1976b. Adapted with permission from Brooke and Melvin: Morphology of Diagnostic Stages of Intestinal Parasites of Man, USDHEW PHS publication no. 1966, 1969.

Dientamoeba fragilis, formerly considered an ameba, has been reclassified as a flagellate, based on electron microscopy studies. *Dientamoeba* is related to the trichomonads, and like *T. hominis* of the intestine and *T. vaginalis* of the vagina, prostate, and urethra, it has a trophozoite stage but lacks a cyst stage.

Three species of *Trichomonas* occur in humans: *T. tenax, T. hominis,* and *T. vaginalis.* All three lack a cyst stage. *T. tenax* is an uncommon flagellate found in the mouth and is associated with poor oral hygiene. *T. hominis* is a pear-shaped organism that lives in the intestinal tract. It has three anteriorly directed flagella and an undulating membrane that runs the length of the body and terminates in a posterior flagellum; it moves in a rapid, jerking manner. It stains poorly with the usual permanent fecal stains and may be easily overlooked. Diagnosis is usually from direct saline mounts of fresh fecal material. Neither *T. hominis* nor *T. tenax* is illustrated in this atlas.

T. vaginalis is responsible for vaginal trichomoniasis. Infections may occur in both women and men, but clinical symptoms are more frequent in women. Transmission is mainly by sexual contact and, less commonly, by contaminated objects. Diagnosis usually is made by examining direct mounts of vaginal discharges or prostatic secretions. As in other trichomonads, the characteristic movement is a diagnostic feature. *T. vaginalis* is illustrated on the cover of this book.

With the exception of *Dientamoeba,* all living trophozoites of the flagellates have characteristic rapid motility that may allow for a specific diagnosis based on this feature alone. Flagella stain poorly — if at all — in routine fecal smear preparations, so other morphologic features are used for specific identification. Shape of the organism, number and configuration of nuclei, and the presence of such structures as a cytostome, sucking disk, fibrils, and undulating membrane, are most useful for identification. *Dientamoeba* trophozoites may have one nucleus, but usually have two nuclei with no peripheral chromatin. The karyosome usually is fragmented into four to eight chromatin granules in a central mass.

Intestinal flagellates illustrated in this atlas include: *Dientamoeba fragilis, Giardia lamblia, Chilomastix mesnili,* and *Trichomonas vaginalis* (cover).

BLOOD AND TISSUE FLAGELLATES. — The flagellate protozoan parasites of blood and tissue comprise species belonging to the genera *Trypanosoma* and *Leishmania.* These organisms have a life cycle involving a bloodsucking, arthropod intermediate host and humans or other mammals as the definitive hosts. These parasites have two or more developmental stages in their life cycles, each of which is characterized morphologically by differences in the site of origin of the flagellum (Fig 2, Table 5).

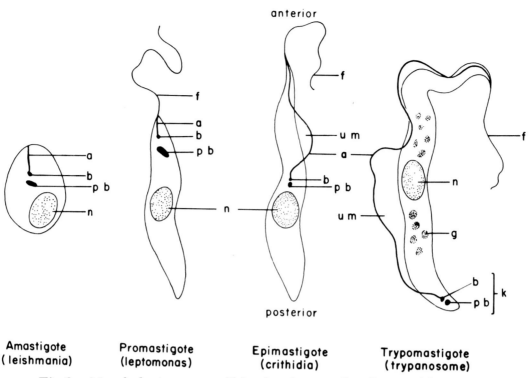

anterior

posterior

| Amastigote
(leishmania) | Promastigote
(leptomonas) | Epimastigote
(crithidia) | Trypomastigote
(trypanosome) |

Fig 2. Morphologic stages of blood and tissue flagellates. Abbreviations: *a*, axoneme; *b*, blepharoplast; *f*, flagellum; *g*, granules; *k*, kinetoplast; *n*, nucleus; *pb*, parabasal body; *um*, undulating membrane. From Smith et al, 1976a. Used by permission.

Amastigote stages of the species of *Leishmania* causing cutaneous and mucocutaneous disease occur in the skin and mucous membranes. In *Leishmania donovani* infections, amastigote stages occur in the viscera. In the case of *Trypanosoma cruzi* infections, amastigotes may be found in reticuloendothelial tissues, glial cells, and in striated muscle. Amastigotes frequently are referred to as Leishman-Donovan (LD) bodies. These intracellular bodies are round or oval-shaped, with a postcentrally located nucleus, and a blepharoplast and axoneme anterior to the nucleus. No external flagellum is present in this stage.

Promastigote stages occurring in *Leishmania* are found in the midgut of the insect vector, where they undergo multiplication and become the infective stage for the mammalian host. This stage also occurs in various culture media following inoculation with amastigotes present in tissue. This stage is elongate and slender, with a central nucleus. The blepharoplast is near the anterior end, from which a short, free flagellum arises.

The epimastigote stage, present only in trypanosomes, is found in either the salivary glands or the midgut of the insect vector, or it also may be grown in culture. This stage is long and slender, with the nucleus postcentrally located, and the blepharoplast immediately anterior to the nucleus. A flagellum arises from the blepharoplast and is directed anteriorly to exit as a free flagellum.

The trypomastigote stage lives extracellularly and occurs only in the trypanosomes. It usually is long and slender, and has a centrally placed nucleus. The blepharoplast giving rise to the flagellum is behind the nucleus, close to the posterior end. The flagellum is directed anteriorly, attached to the body by an undulating membrane, and exits at the anterior end as a long, free flagellum.

Leishmaniasis is diagnosed by detecting the amastigote stage in skin biopsy specimens or impression smears, or by culture techniques in which suspect infected tissue is inoculated into culture media, producing promastigote forms.

African trypanosomiasis (*Trypanosoma gambiense* or *T. rhodesiense*) is diagnosed by detecting trypomastigotes in lymph node aspirates, in blood, or in cerebrospinal fluid. *T. cruzi* has both a tissue stage (amastigotes occurring in nests in reticuloendothelial cells, myocardium, and other tissues) and a bloodstream stage in which trypomastigotes are found.

Blood and tissue flagellates illustrated in this atlas include: *Leishmania* spp, *Trypanosoma gambiense* or *T. rhodesiense*, and *T. cruzi*.

TABLE 5.—Blood and Tissue Flagellates Found in Humans*

SPECIES	DEVELOPMENTAL STAGES				TRANSMISSION	VECTORS
	AMASTIGOTE	PROMASTIGOTE	EPIMASTIGOTE	TRYPOMASTIGOTE		
Leishmania						
L. donovani	Intracellular, in reticuloendothelial system, lymph nodes, liver, spleen, bone marrow, etc. Culture.	Midgut and pharynx of vector. Culture.	–	–	Bite.	Sand flies (*Phlebotomus, Lutzomyia*).
L. tropica and L. braziliensis	Intracellular and extracellular in skin and mucous membranes of humans.	Midgut and pharynx of vector. Culture.	–	–	Bite.	Sand flies (*Phlebotomus, Lutzomyia*).
Trypanosoma						
T. gambiense	–	–	Salivary glands of vector. Culture.	Blood, lymph nodes, cerebrospinal fluid of final host. Intestine and salivary gland of vector.	Bite.	Tsetse fly (*Glossina*).
T. rhodesiense	–	–	Salivary glands of vector. Culture.	Blood, lymph nodes, cerebrospinal fluid of final host. Intestine and salivary glands of vector.	Bite.	Tsetse fly (*Glossina*).
T. cruzi	Intracellular in viscera, myocardium, brain of humans. Tissue culture.	Intracellular, but transitional, in humans.	Midgut of vector. Culture.	Blood (temporary) of man. Intestine and rectum, feces of vector. Culture.	Feces of vector into wound.	Reduvid bugs (Triatominae).
T. rangeli	–	Transitional only.	Midgut of vector. Culture.	Blood of humans. In hemolymph, salivary glands, and proboscis of vector. Culture.	Bite.	Reduvid bugs (Triatominae).

*From Smith et al, 1976a. Used by permission.

CILIATES

The only ciliate parasite of humans is *Balantidium coli*. A common parasite of pigs, *Balantidium* has both trophozoite and cyst stages. It lives in the colon where it may produce severe ulcerations. The morphologic features of this parasite are presented in the description accompanying Plate 19.

COCCIDIA

The intestinal and tissue-dwelling coccidian parasites of humans have complex life cycles in which both sexual and asexual reproduction may occur, and intermediate hosts may or may not be used. Human infections may be caused by one or more species of the genera *Isospora*, *Sarcocystis*, and *Toxoplasma*. Considerable controversy exists over the classification and specific identification of many of these parasites so that taxonomic designations now used may change in the future.

Isospora belli does not commonly occur in the human intestine where infections may be asymptomatic or may be associated with diarrhea of varying duration. Infections are diagnosed by finding immature oocysts in feces.

Species of *Sarcocystis* infect a wide range of hosts including humans, and have an obligatory two-host (predator-prey) life cycle. Humans may serve as dead-end, intermediate hosts for many species. These infections usually are detected by histologic demonstration of cysts in muscle tissue. Human intestinal infections may be caused by several species for which pigs, cattle, and other animals ordinarily serve as intermediate hosts. Infection is acquired by ingestion of raw or poorly cooked meat containing the cyst stages.

These infections are diagnosed by finding mature oocysts containing sporocysts or individual sporocysts in feces. Although many species of *Sarcocystis* have been described, the size and structure of the oocysts and sporocysts found in feces are so similar, overlapping to such an extent, that species diagnoses are difficult to make. A parasite of humans known as *Isospora hominis* in the old literature now is recognized as belonging to the genus *Sarcocystis*.

Toxoplasma gondii is a coccidian parasite known to be an important disease agent in newborns, adults, and immunosuppressed persons. Members of the cat family, both wild and domesticated, are the only hosts in which the intestinal phase of the life cycle occurs. In these instances, small, immature oocysts, approximately $13 \times 11 \mu$, are passed sporadically in feces. A wide range of hosts, including birds and mammals as well as humans, may serve as intermediate hosts. Human infection may be acquired transplacentally, by ingestion of sporulated oocysts, or by ingestion of tissue cysts present in poorly cooked meat or fowl. In humans, these cyst stages may contain many organisms (zooites). They can occur in individual cells (macrophages or leukocytes), or in tissues (brain and muscle). Diagnosis of human infection is based mainly on clinical symptoms, serologic diagnosis, or by injection of infected tissues into laboratory mice. *Toxoplasma* organisms sometimes are detected by impression smears of lymph nodes, bone marrow, spleen, and brain and tissue exudates.

Coccidian organisms illustrated in this atlas include: *Isospora belli*, *Sarcocystis bovicanis*, and *Toxoplasma gondii*.

Diagnosis of Helminthic Infections

Helminth infections usually are diagnosed by finding the characteristic eggs, larvae, or proglottids in feces. Most of the helminths living in the intestinal tract or in associated organs produce eggs that are passed in feces. Each parasitic species produces eggs that exhibit natural biologic variation in size, but are, for the most part, highly uniform in shape, coloration, and stage of development. Criteria for identification of eggs are:

Size.—Helminth eggs range in length from small (25μ) to large $(150 \mu$ or longer), and have diameters as small as $12-14 \mu$ for some of the small trematode eggs to 90μ or more for the larger trematode eggs. Comparative sizes of the more common helminth eggs are illustrated in Figure 3.

Shape.—The shape of eggs, varying from spherical to elongate, is a constant feature for most species of helminths. Exceptions do occur, however. For example, in *Ascaris lumbricoides*, infertile eggs are more elongate than are fertile eggs, and in *Trichuris trichiura*, eggs may develop in bizarre shapes following chemotherapy. Subtle differences in shape may be significant in separating species within a particular genus. In *T. vulpis*, eggs tend to be more barrel-shaped than those of *T. trichiura*, whereas the eggs of *Opisthorchis viverrini* frequently are broader than those of the closely related *Clonorchis sinensis*.

Eggshell.—Helminth eggs usually have a smooth shell that may vary considerably in thickness, depending upon the species. With the exception of *A. lumbricoides* eggs, which have a mammillated external shell layer, the smoothness of the shell often is an important feature in separating eggs from vegetable cells or other plant material that may have irregular, limiting membranes.

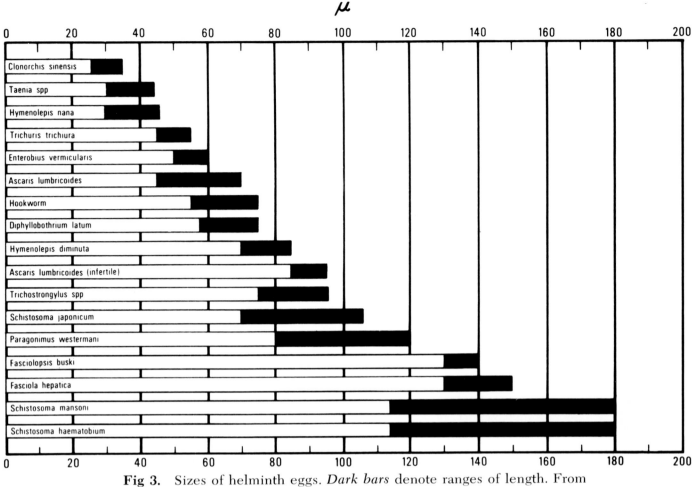

Fig 3. Sizes of helminth eggs. *Dark bars* denote ranges of length. From Smith et al, 1976c. Used by permission.

The color of the shell generally becomes more apparent as its thickness increases, and it may vary from colorless to yellowish-brown. Examples of thick-shelled eggs are those of *Enterobius*, which are colorless, and *Ascaris* and *Trichuris*, which usually are bile-stained. Hookworms and trichostrongyles have thin-shelled eggs that lack color. Infertile eggs, such as those of *Ascaris*, have a thinner shell than their fertile counterparts.

Various modifications of shell structure also are important in identification. These include clear, mucoid polar "plugs" (*Trichuris* and *Capillaria*), mammillations (*Ascaris*), pitting (*Toxocara*), striations (*Taenia* and *Capillaria*), spines (schistosomes), knobs (*Diphyllobothrium* and *Clonorchis*), and an operculum, as in *Diphyllobothrium* and all trematode eggs other than the schistosomes.

Stage of development.—The stage of the ovum's development within freshly passed eggs is characteristic for each parasite species. Most nematode eggs are unembryonated when passed in feces. When stool specimens are allowed to stand at warm temperatures for a day or longer prior to their exam-

ination, eggs may develop, embryonate, and even hatch (hookworm eggs in particular), and thus present diagnostic problems.

In *Strongyloides*, a first-stage larva is the stage normally excreted in feces. Most human tapeworm infections are diagnosed by finding eggs that contain six-hooked embryos (oncospheres) in feces. Eggs of *Diphyllobothrium* are an exception in that they are unembryonated when passed. Trematode eggs are either embryonated, contain a miracidium (schistosomes and the flukes that produce small eggs), or are unembryonated at time of passage (those species with eggs larger than 60 μ). Infertile eggs are characterized by an unorganized, granular mass including fat globules and refractile material.

NEMATODES

Nematodes typically have five stages in their life cycle—four larval and the adult. Many intestinal nematodes have a direct life cycle in which infective stages are found primarily in the soil. Filarial worms all require an arthropod vector as an intermediate host for transmission of infection. *Trichinella spiralis* infection is acquired by ingestion of flesh containing infective larvae.

Enterobius vermicularis probably is the most common nematode infection of humans because its eggs rapidly develop to the infective stage, and are readily transmitted from one infected person to others by fomites. Many soil-transmitted nematodes (*Ascaris, Trichuris,* hookworms; *Strongyloides, Trichostrongylus, Capillaria,* and *Toxocara*) have eggs or larval stages that require development in the external environment before becoming infective. Hence, the prevalence of these parasites in any geographical area is dependent upon soil type, climate, level of sanitation in the community, and the behavioral characteristics of the population.

The diagnosis of most intestinal nematode infections depends on finding and identifying the characteristic eggs in feces. The identification of nematode larvae in fecal specimens, usually limited to *S. stercoralis,* should not be difficult if fecal specimens are examined promptly, and if plant hairs or fibers are not confused with larvae.

Nematode larvae have a tapering, rounded anterior end, and a posterior extremity that may terminate bluntly or with a pointed end. An alimentary tract consists of an esophagus that opens to the intestine and terminates in a ventrally situated anus near the posterior end. Between the oral opening and the anterior end of the esophagus may be a buccal canal that varies greatly in length. This can be of diagnostic significance, for example, in separating first-stage hookworm and *Strongyloides* larvae. A genital primordium, a cluster of cells that gives rise to the adult reproductive system, may or may not be evident in some larvae. Usually, the genital primordium lies approximately at the level of midintestine, between the intestine and the ventral body wall.

Strongyloides first-stage larvae normally are the only nematode larvae that will be encountered in examining fresh fecal specimens. Hookworm first-stage larvae are found only in fecal specimens that have been allowed to stand at room temperature 24–36 hours prior to examination. Differential features of the hookworm and *Strongyloides* larvae are illustrated in Figure 4 and in Plate 35.

The differentiation of nematode larvae from plant hairs should not be difficult. Although plant hairs may have one end that is tapered and rounded, the opposite end usually will appear blunt and jagged. In addition, the refractile central canal of these hairs is unlike the nematode alimentary tract, which is clearly differentiated into an esophagus and intestine.

Filarial worms are long, thread-like nematodes

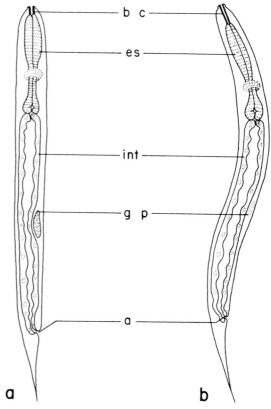

Fig 4. Morphology of first-stage rhabditoid larvae: *Strongyloides* (a) and hookworm (b). The characteristic morphologic features useful in differentiating these larvae are shown here. Abbreviations: *a,* anus; *bc,* buccal canal; *es,* esophagus; *gp,* genital primordium; *int,* intestine.

that, as adults, live in body cavities, the subcutaneous tissues, or the lymphatic system of the human host (Table 6). Female worms produce embryos, called *microfilariae,* that are released into the bloodstream or dermal layers of the skin, depending on the species involved. Bloodsucking arthropods ingest microfilariae. In these intermediate hosts, over a period of 7–14 days, the microfilaria develops to the infective, third stage that is transmitted to the next host when the arthropod feeds again. Filarial infections usually are diagnosed by detecting microfilariae in blood or in skin. The specific identification of microfilariae is based on their structure, their location in the host (blood or skin), and their periodicity (time of day in which they usually occur in blood).

Depending on the species, microfilariae usually measure 150–350 μ in length, with a maximum diameter of 4–10 μ. Microfilariae measured in stained blood films usually are smaller than those measured in formalin-fixed specimens. They are vermiform and have a simple cellular organization. Some species retain the delicate vitelline mem-

TABLE 6.—Common Human Filarial Parasites*

CHARACTERISTIC	Wuchereria bancrofti	Brugia malayi	Loa loa	Mansonella ozzardi	Dipetalonema perstans	Dipetalonema streptocerca	Onchocerca volvulus
Geographic distribution	Cosmopolitan; tropics and subtropics	Asia, Indian subcontinent	West and Central Africa	South and Central America, Caribbean	Africa, South and Central America	West and Central Africa	Africa, Central and South America
Adult habitat	Lymphatic system	Lymphatic system	Subcutaneous tissues	Mesenteries, body cavities (?)	Body cavities, mesenteries, perirenal and retroperitoneal tissues (?)	Subcutaneous tissues	Subcutaneous tissues
Vector	Mosquitoes	Mosquitoes	*Chrysops* (deer fly)	*Culicoides, Simulium*	*Culicoides* (midge)	*Culicoides* (midge)	*Simulium* (black fly)
Location of microfilariae	Blood	Blood	Blood	Blood	Blood	Skin	Skin
Periodicity	Nocturnal†	Nocturnal‡	Diurnal	None	None	None	None
Morphology of microfilariae:							
Sheath	Present	Present	Present	Absent	Absent	Absent	Absent
Length (μ) Smears	244–296 (260)	177–230 (220)	231–250 (238)	163–203 (183)	190–200 (195)		
2% Formalin	275–317 (298)	240–298 (270)	270–300 (281)	203–254 (224)	183–225 (203)	180–240 (210)	304–315 (309)
Skin snips	–	–	–			–	–
Width (μ)	7.5–10.0	5–6	5–7	3–5	4–5	5–6	5–9
Tail and tail nuclei	Tapered to point; no nuclei in end of tail	Tapered; terminal and subterminal nuclei	Tapered; nuclei irregularly spaced to end of tail	Long, slender tail; no nuclei in end of tail	Tapered, bluntly rounded; nuclei to end of tail	Tapered, bluntly rounded; nuclei to end of tail. Tail bent in hook shape	Tapered to point; no nuclei in end of tail

*From Smith et al. 1976a. Used by permission.
†Subperiodic in Pacific Islands.
‡Subperiodic form as well.

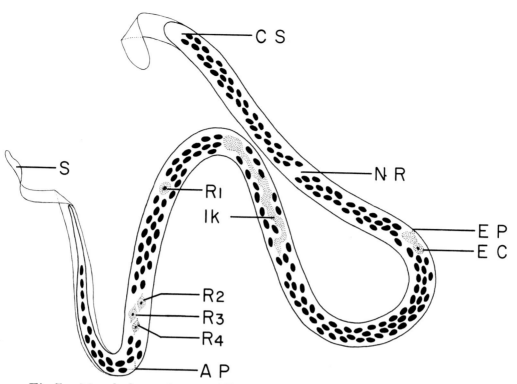

Fig 5. Morphology of a microfilaria. This representation shows character-istic anatomic features, including the sheath *(S)*, cephalic space *(CS)*, nerve ring *(NR)*, excretory pore *(EP)*, excretory cell *(EC)*, Innenkörper *(Ik)*, rectal cells *(R1–R4)*, and anal pore *(AP)*. Modified from Smith et al, 1976a. Used by permission.

brane as a sheath that adheres to the body of the microfilaria along its length, but may project beyond the body at either or both ends.

In stained specimens, the microfilaria consists of a column of nuclei separated at various intervals by specific anatomic reference points (Fig 5). For identification, the key diagnostic features include the cephalic space, excretory pore and cells, the presence or absence of an Innenkörper, the number and distribution of nuclei in the tail, the rectal cells, and the presence or absence of the sheath (Fig 6). The area occupied by the Innenkörper has a sparse number of nuclei, stains lightly, and is found in the midbody region of some microfilariae.

In addition to the usual nematode infections, a number of animal parasites produce zoonotic infections in humans. These include dog and cat ascarids (*Toxocara* spp, which cause visceral larva migrans), *Capillaria* species from rodents and birds, and various other genera, including *Dioctophyma, Oesophagostomum, Physaloptera*, and *Gongylonema*.

The nematodes illustrated in this atlas include: *Enterobius vermicularis, Ascaris lumbricoides, Trichuris trichiura, T. vulpis, Capillaria hepatica, C. philippinensis*, hookworms, *Strongyloides stercoralis, Trichostrongylus* sp, *Trichinella spiralis,*

Physaloptera sp, *Gongylonema* sp, *Oesophagostomum* sp, *Dioctophyma renale, Toxocara canis, T. cati, Toxascaris leonina, Wuchereria bancrofti, Brugia malayi, B. timori, Loa loa, Mansonella ozzardi, Dipetalonema perstans, D. semiclarum, D. streptocerca*, and *Onchocerca volvulus*.

TREMATODES

Trematodes (flukes) have complex life cycles involving at least two hosts. Various snails serve as obligatory first intermediate hosts. They are infected by the miracidium, a ciliated larval stage, that develops within the fluke egg. Small fluke eggs (those less than 40–50 μ) and the eggs of the schistosomes contain a miracidium when passed, whereas the larger trematode eggs (more than 50 μ long) are unembryonated when passed, and must undergo a period of development in water before the miracidium is produced.

With the exception of the schistosomes, all trematode eggs have an operculum that aids in their identification. The operculum is easy to see in some species (*Clonorchis, Paragonimus*), whereas it may be difficult to detect in others (*Fasciola, Fasciolopsis*, heterophyids). In addition to the operculum,

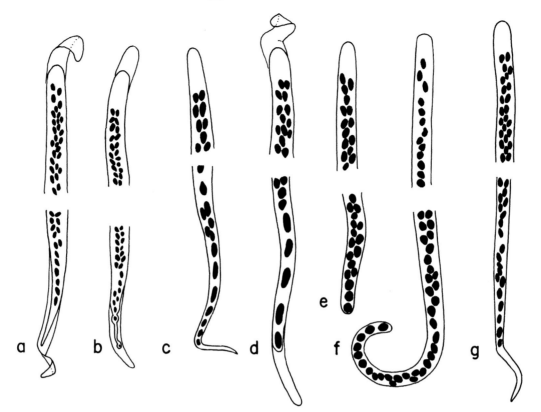

Fig 6. Anterior and posterior ends of microfilariae found in humans: *Wuchereria bancrofti* (**a**), *Brugia malayi* (**b**), *Onchocerca volvulus* (**c**), *Loa loa* (**d**), *Dipetalonema perstans* (**e**), *D. streptocerca* (**f**), and *Mansonella ozzardi* (**g**). Modified from Smith et al, 1976a. Used by permission.

the shell of fluke eggs may be thickened *(Paragonimus)* or have a knob or projection *(Clonorchis)* at the abopercular end. Schistosome eggs have characteristic spines on the shell, although these may be inconspicuous in some species *(Schistosoma japonicum and S. mekongi)*.

Trematode species illustrated in this atlas include: *Heterophyes heterophyes, Metagonimus yokogawai, Echinostoma ilocanum, Fasciolopsis buski, Fasciola hepatica, Clonorchis sinensis, Opisthorchis viverrini, Dicrocoelium dendriticum, Paragonimus westermani, Schistosoma mansoni, S. japonicum, S. mekongi,* and *S. haematobium.*

CESTODES

Both larval and adult cestodes (tapeworms) may infect humans. Larval infections, such as cysticercosis or hydatid, usually are diagnosed by serologic findings or detection of the parasite in tissues. In humans, all adult tapeworms live in the small intestine, and infections are diagnosed by finding eggs or proglottids or both in feces. Pets in the domestic environment may be a source of uncommon human

tapeworm infections. Two examples of these, *Bertiella* and *Mesocestoides*, are illustrated in Plate 73.

Diphyllobothrium latum, the broad fish tapeworm, is distinguished from the other human tapeworms by its operculate, unembryonated eggs. Although sometimes confused with lung fluke *(Paragonimus)* eggs, the eggs of *D. latum* are smaller, and usually have a knob at their abopercular end. All the other major human tapeworm species produce eggs that contain a six-hooked embryo (oncosphere) when passed in feces.

Taenia solium and *T. saginata* infections may be diagnosed by finding eggs or proglottids in feces. Use extreme caution in handling *Taenia* proglottids since the eggs of *T. solium* (but not *T. saginata*) are infective to humans, and, when ingested, may produce cysticercosis. The eggs of any *Taenia* species infecting humans or animals, as well as those of the genus *Echinococcus* are identical and cannot be differentiated from one another.

Cestode species illustrated in this atlas include: *Diphyllobothrium latum, Taenia saginata, T. solium, Hymenolepis nana, H. diminuta, Dipylidium caninum, Echinococcus granulosus, Bertiella* sp, and *Mesocestoides* sp.

TABLE 7.—Nonparasitic Objects[a]

ARTIFACT	RESEMBLANCE	SALINE MOUNT	DIFFERENTIAL CHARACTERISTICS OF ARTIFACT — PERMANENT STAIN	
			CYTOPLASM	NUCLEUS
Polymorphonuclear leukocytes; seen in dysentery and other inflammatory bowel diseases. Have mean diameter of 14 μ.	E. histolytica cyst	May be a problem. Granules in cytoplasm. Cell border irregular.	Less dense, often frothy. Border less clearly demarcated than that of ameba.	More coarse. Larger, relative to size of organism. Irregular shape and size. Chromatin unevenly distributed. Chromatin strands may link nuclei.
Macrophages; seen in dysentery and other inflammatory bowel diseases. May be present in purged specimens. Measure 30–60 μ.	Amebic trophozoite, especially E. histolytica	Nuclei larger and of irregular shape, with irregular chromatin distribution. Cytoplasm granular; may contain ingested debris. Cell border irregular and indistinct. Movement irregular and pseudopodia indistinct.	Coarse. May contain red blood cells and other inclusions.	Large and often irregular in shape. Chromatin irregularly distributed.
Squamous epithelial cells (from anal mucosa).	Amebic trophozoite	Nucleus refractile and large. Cytoplasm smooth. Cell border distinct.	Stains poorly.	Large and single. Large chromatin mass may resemble karyosome.
Columnar epithelial cells (from intestinal mucosa).	Amebic trophozoite	Nucleus refractile and large. Cytoplasm smooth. Cell border distinct.	Stains poorly.	Large with heavy chromatin on nuclear membrane. Often large central chromatin mass resembling karyosome.
Blastocystis hominis, yeast-like organism that frequently grows in feces; ruptures in water. May be 5–30 μ.	Protozoan cyst	Spherical to oval. 6–15 μ in length. Central clear area. Peripheral refractile granules (3–7) may resemble nuclei.	Central mass may stain light or dark. Prominent wall.	Peripheral granules may resemble nuclei. Granules vary in size and appearance. True nuclear structure not present.
Yeasts (Normal constituent of feces). Usually 4–6 μ.	Protozoan cyst	Oval. Thick wall. No internal structure. Budding forms may be seen.	Oval-shaped. Little internal structure. Refractile cell wall. Budding forms may be seen.	None.
Starch granules. May vary considerably in size.	Protozoan cyst	Rounded or angular. Very refractile. No internal structure. Stain pink to purple in iodine mounts.	Not a problem in permanently stained slides.	

Note: Other artifacts, such as contaminating plant cells and pollen grains, are occasionally seen. These should not be difficult to differentiate.
[a]From Smith et al, 1976b. Used by permission.

Artifacts

A major problem in diagnostic parasitology is the ability to distinguish parasitic organisms from those elements in feces and blood that resemble them. The term *artifact* is used for objects that may be confused with parasites. Fecal material normally consists of food residue, various products of digestion, sloughed epithelial cells, mucus, and microorganisms, such as bacteria and yeasts. Hence, it is not surprising that this bewildering array of material contributes to frequent errors in identification of trophozoites, cysts, eggs, or larvae (Table 7).

Both wet mount preparations and permanently stained fecal smears may contain a variety of cellular elements and plant material that resemble protozoan organisms, and helminth eggs and larvae. Many cells of human origin, including macrophages, polymorphonuclear leukocytes, and epithelial cells, may be found in feces and pose diagnostic difficulties. Plant hairs, such as the fuzz on peaches, may be misidentified as larvae because they superficially resemble nematode larvae in size and shape. Many kinds of pollen grains and other plant cells are found in feces, and some of these have a remarkable resemblance to *Taenia* eggs and other helminth eggs. Other objects that may occur in feces as a result of accidental ingestion or contamination of the fecal specimen include mite eggs and the eggs and larvae of plant nematodes, such as *Meloidogyne (Heterodera)*.

Spurious infections occur when parasite eggs accidentally are ingested, pass through the intestinal tract, and are excreted in feces for a short (several days to a week) period of time. For example, eating cattle liver infected with *Fasciola hepatica* or *Dicrocoelium dendriticum* may result in excretion of these eggs digested from the liver without actual presence of human infection. Regional food habits, such as eating squirrel livers, may result in the presence of eggs of *Capillaria hepatica*, a rodent parasite, in feces. Other more exotic—and frequently unidentifiable—eggs may occur in feces, representing spurious infection acquired by eating the flesh of fish, birds, or other vertebrate or invertebrate hosts.

Artifacts simulating parasites in blood films include fibers that become incorporated into these films prior to staining and may resemble microfilariae to the inexperienced observer. Failure to rinse blood films properly following staining may result in precipitates of stain on red blood cells that may be confused with malarial organisms. In general, artifacts in blood films lack symmetry, stain irregularly, and lack clearly defined structure.

Artifacts and fecal elements illustrated in this atlas include: plant hairs and cells, pollen grains, diatoms, Charcot-Leyden crystals, mite eggs, Beaver bodies, plant nematode eggs, and *Blastocystis* and yeasts.

Part II

Atlas

Entamoeba histolytica

CLASSIFICATION. — Ameba.

DISEASE. — Amebiasis.

GEOGRAPHIC DISTRIBUTION. — Cosmopolitan.

LOCATION IN HOST. — Lumen of colon and cecum; tissue invasion may involve colon, and extraintestinal locations including liver, lung, brain, skin, and other tissues.

MORPHOLOGY. — Trophozoites. — May range in size from $10-60 \mu$; noninvasive forms usually are $15-20 \mu$, and invasive organisms are more than 20μ. Living trophozoites exhibit progressive, sometimes explosive, motility with extrusion of hyaline, finger-like pseudopodia. The single nucleus is not visible in unstained preparations. With stain, the nucleus has a small, compact karyosome that usually is centrally located but may be placed eccentrically. Peripheral chromatin typically is finely granular and is distributed evenly over the inner surface of the nuclear membrane, although occasionally its deposition may be irregular. The cytoplasm usually is finely granular. Noninvasive amebae may contain bacteria and invasive forms may take in red blood cells. The presence of erythrocytes in the cytoplasm of trophozoites is considered diagnostic for *E. histolytica*. The cytoplasm may undergo considerable vacuolation when preservation of the specimen is delayed.

Cysts. — Mature cysts are spherical and contain four nuclei. Although these cysts may range from $10-20 \mu$, the usual size is $12-15 \mu$. Nuclei are invisible in unstained preparations. Immature cysts may contain one or two nuclei. Except for their smaller size, the structure of nuclei in cysts is similar to the nuclei in trophozoites. The glycogen present in mature cysts usually is diffuse, but it may be concentrated into a discrete mass in young cysts. Chromatoid bodies, when present, generally are elongated bars with bluntly rounded ends.

LIFE CYCLE. — Direct transmission by ingestion of cyst stage.

DIAGNOSIS. — Intestinal amebiasis is detected by presence of trophozoite or cyst stages in feces, in proctoscopic material, or other aspirates. Extraintestinal amebiasis may be diagnosed on the basis of clinical history, use of serologic tests, liver scans, and demonstration of organisms in histologic specimens.

Diagnostic problems. — Nuclear structure may exhibit considerable variation from the usual, small, central karyosome with fine, evenly distributed peripheral chromatin. Not uncommonly, the karyosome may be eccentric and the peripheral chromatin may be distributed unevenly. In these instances, the organisms may be confused with *E. coli*. In trophozoites undergoing degeneration, the nucleus usually is the first structure to change; this may include fragmentation of the karyosome and/or alteration of the peripheral chromatin. Young cysts containing discrete masses of glycogen may be confused with cysts of *Iodamoeba*, however, in such instances examination of the nuclear structure will lead to the diagnosis of *E. histolytica*.

FIGURES. — Plate 1:1-5; Plate 2:1-6; Plate 3:1-6.

PLATE 1

Entamoeba histolytica, Trophozoites, Trichrome Stain

Fig 1. Note the typical nucleus and vacuolated cytoplasm in this trophozoite. Vacuolation of cytoplasm sometimes is seen in organisms in fresh fecal specimens. However, this occurs more frequently in organisms from fecal samples that have been left standing at room temperature for several hours.

Fig 2. This trophozoite from a dysenteric stool has finely granulated cytoplasm, and contains several red blood cells in varying stages of digestion. Note the large, vacuolated pseudopod at the lower right margin. The nucleus in this organism contains a typical, small, discrete, central karyosome.

Fig 3. Note this ameba's characteristic nuclear structure, ie, a discrete, small karyosome, and evenly distributed peripheral chromatin.

Fig 4. Another typical trophozoite containing red blood cells, also from a dysenteric stool.

Fig 5. Trophozoite with characteristic nucleus and cytoplasm.

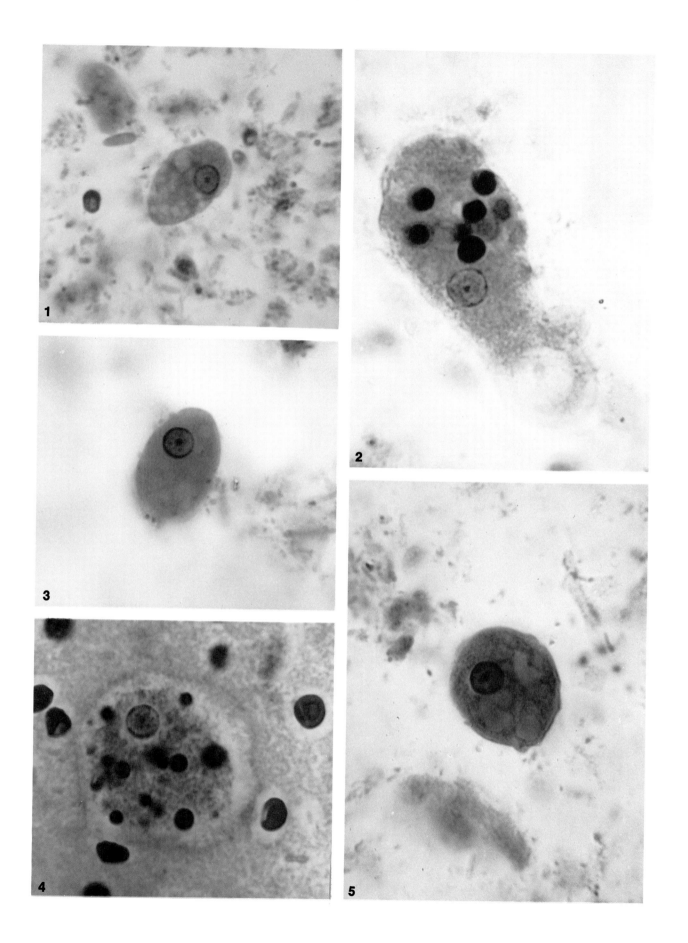

PLATE 2

Entamoeba histolytica, Cysts

Fig 1.   An iodine-stained, mature cyst with four nuclei.

Fig 2.   A uninucleate, iodine-stained cyst in formalin-preserved feces showing typical nuclear structure and presence of a glycogen mass. With prolonged storage, the staining quality of glycogen may be altered (as in this specimen). Nuclei in immature cysts typically are larger than those in mature cysts.

Fig 3.   Another iodine-stained cyst. This organism is binucleate and contains a chromatoid body and poorly staining glycogen.

Fig 4.   Note the large sizes of both the uninucleate cyst stained with trichrome and its nucleus with a diffuse karyosome. Rounded chromatoid bodies surround a lightly stained glycogen vacuole.

Fig 5.   This binucleate cyst stained in trichrome contains two chromatoid bodies—one large and deeply stained, the other with a less conspicuous size and staining quality. The nuclei, though larger than usually seen in a mature cyst, demonstrate characteristic morphologic features.

Fig 6.   A trophozoite and a mature cyst containing chromatoid bodies are seen in the same field. The nucleus of this trophozoite has a small, eccentric karyosome.

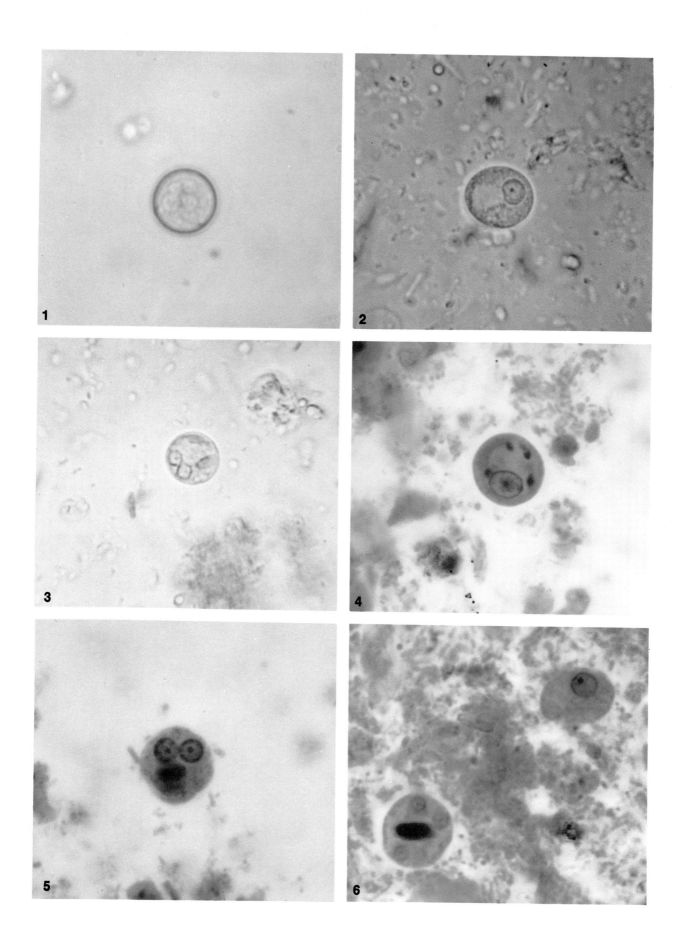

PLATE 3

Entamoeba histolytica, Trophozoites and Cysts

Fig 1. A trophozoite stained with iron hematoxylin shows characteristic nuclear structure. The cytoplasm is finely granular and contains red blood cells.

Fig 2. In this trichrome-stained, uninucleate cyst, note the size of the nucleus with a "spread-out" karyosome, and of the glycogen vacuole surrounded by chromatoid bodies.

Fig 3. This uninucleate cyst stained with iron hematoxylin has the typically large nucleus and a deeply stained chromatoid body.

Fig 4. An immature, binucleate cyst stained with trichrome. The nuclei are large; one is less conspicuous than the other because of the plane of focus. Two chromatoid bodies are present.

Fig 5. A mature, iron hematoxylin-stained cyst shows four nuclei and a densely staining chromatoid body in the center of the cyst, all of which are in different focal planes.

Fig 6. Another mature cyst stained with trichrome. The four nuclei are readily identified and no chromatoid bodies are present.

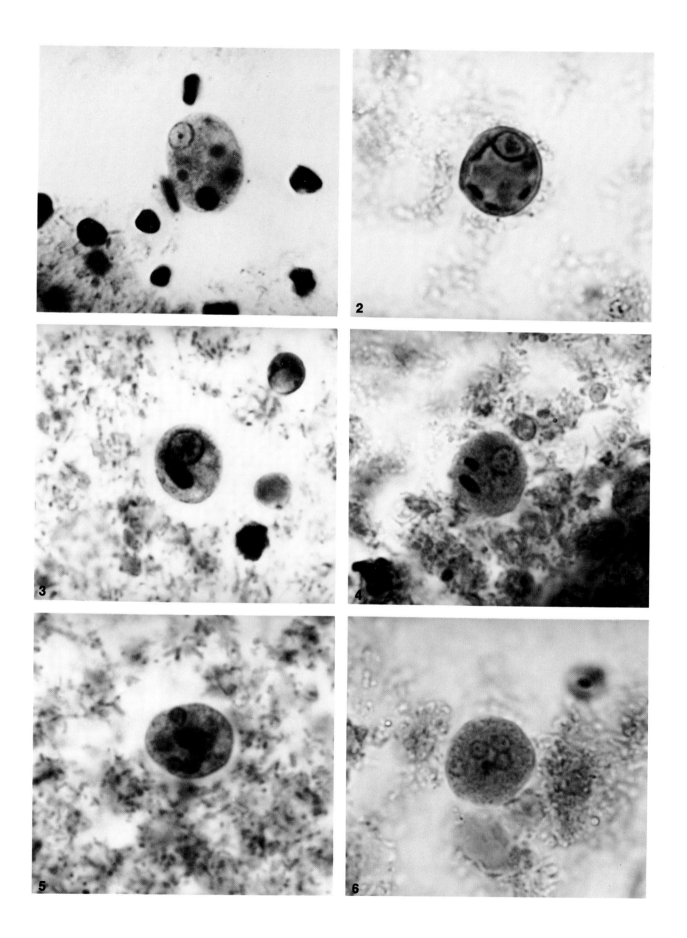

Entamoeba hartmanni

CLASSIFICATION. — Ameba.

DISEASE. — Generally considered to be non-pathogenic; this status is questioned by some workers.

GEOGRAPHIC DISTRIBUTION. — Cosmopolitan.

LOCATION IN HOST. — Lumen of colon and cecum.

MORPHOLOGY. — Trophozoites. — This stage measures 5–12 μ, with an average size of 8–10 μ. The movement of living trophozoites usually is nonprogressive. Trophozoites, as well as cysts, are small, delicate, and usually stain lightly. The single nucleus is not visible in unstained preparations. When stained, it has a small, compact, usually centrally located karyosome, although it is not uncommonly placed eccentrically. Peripheral chromatin generally is present in the form of fine granules of uniform size and distribution. Sometimes the chromatin has a finely beaded appearance. The cytoplasm is finely granular and may contain bacteria, but not red blood cells.

Cysts. — Mature cysts have four nuclei, are spherical, and measure approximately 5–10 μ, with the usual range being 6–8 μ. Nuclei are not visible in unstained wet mounts but they do stain with iodine.

Immature cysts with one or two nuclei are found more commonly than those with four nuclei. In stained preparations, the nuclei usually have a small, discrete, centrally located karyosome, and peripheral chromatin is evenly distributed as fine, uniform granules. As in *E. histolytica,* glycogen may be diffuse in mature cysts but may be more discrete in immature cysts. Chromatoid bodies generally are elongate with bluntly rounded ends.

LIFE CYCLE. — Transmission is direct by ingestion of the cyst stage.

DIAGNOSIS. — Demonstration of trophozoites or cysts in feces.

Diagnostic problems. — This species once was considered to be the "small race" of *E. histolytica.* Although *E. hartmanni* now is recognized as a distinct species, its morphology is similar to that of *E. histolytica* except that the former is smaller. There is some difficulty in separating trophozoites and cysts of *E. hartmanni* and of *E. histolytica* since their sizes may overlap at the respective upper and lower size ranges of the two species. The karyosome of *E. hartmanni* usually is smaller and more compact than that of *E. histolytica,* and *E. hartmanni* generally takes a more delicate stain than does *E. histolytica.*

FIGURES. — Plate 4:1–6; Plate 5:1–6.

PLATE 4

Entamoeba hartmanni, Trophozoites, Trichrome Stain

Figs 1–6. Typical *E. hartmanni* trophozoites are small, and characteristically take a light stain. As outlined in the preceding description the nucleus is small, has a discrete central karyosome, and fine peripheral chromatin. The cytoplasm, which usually is finely granular, never contains red blood cells.

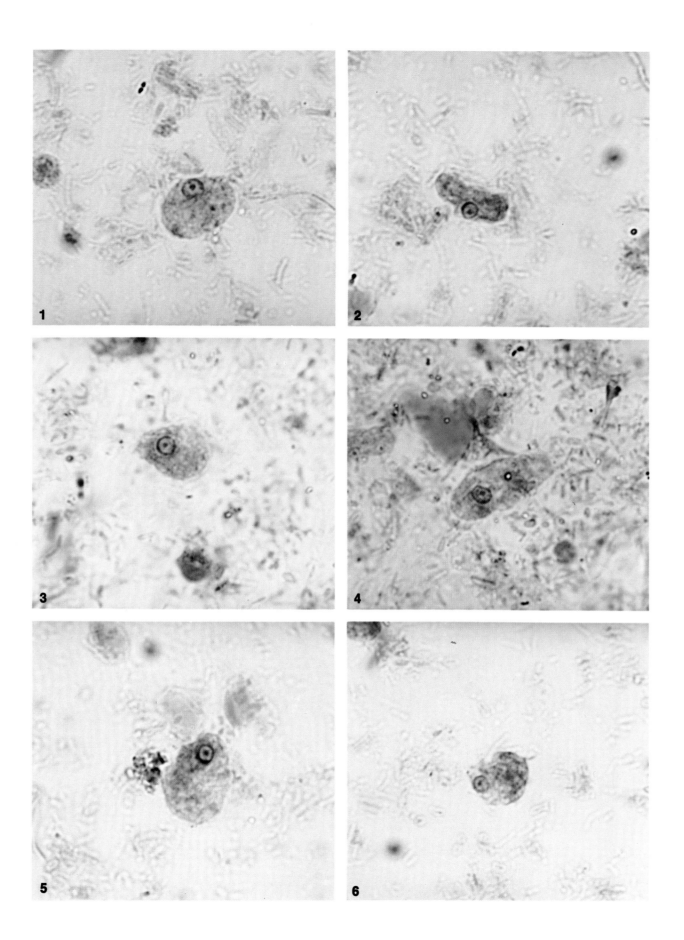

PLATE 5

Entamoeba hartmanni, Trophozoites and Cysts

Fig 1. In this iodine-stained, binucleate cyst, note that the two nuclei (though not in sharp focus) contain discrete, punctate karyosomes. Although these cysts are smaller than those of *E. histolytica,* the diagnosis must be confirmed by examining permanently stained fecal smears.

Fig 2. This uninucleate cyst stained with trichrome has a characteristic nucleus and a small chromatoid body.

Fig 3. In this trichrome-stained mature cyst, only one small nucleus is visible in this plane of focus. The large chromatoid body has typical rounded ends.

Fig 4. A binucleate, trichrome-stained cyst with two chromatoid bodies.

Fig 5. A typical trophozoite stained with trichrome.

Fig 6. In this field one can see a trophozoite of *E. hartmanni* (lower left), and a trophozoite of *Endolimax nana* (upper right). Note the similar size of the two species and the difference in nuclear structure. As mentioned earlier, *E. hartmanni* nuclei have discrete, central karyosomes with fine peripheral chromatin, whereas *E. nana* nuclei have a large karyosome and no peripheral chromatin.

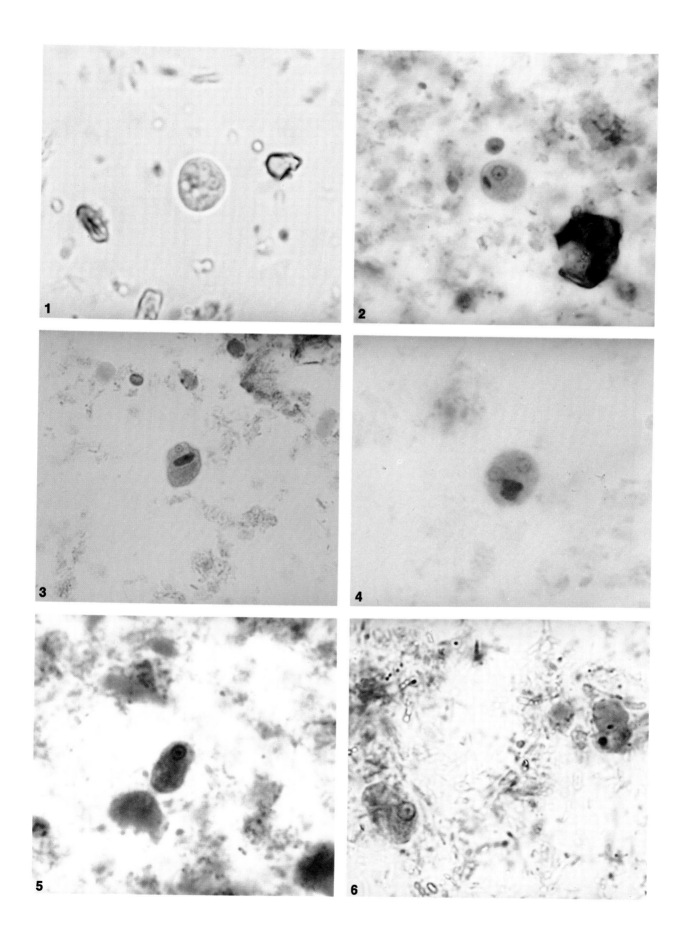

Entamoeba coli

CLASSIFICATION. — Ameba.

DISEASE. — Nonpathogenic.

GEOGRAPHIC DISTRIBUTION. — Cosmopolitan.

LOCATION IN HOST. — Lumen of colon and cecum.

MORPHOLOGY. — Trophozoites. — Although their size range is 15–50 μ, trophozoites usually are 20–25 μ. Living trophozoites typically are sluggish, extrude short, blunt pseudopodia and they exhibit nondirectional movement. The single nucleus often is visible in unstained preparations. In stained organisms, the nucleus has a large, noncompact karyosome that usually is located eccentrically. Peripheral chromatin characteristically is in the form of coarse granules, irregular in size and distribution on the nuclear membrane. It occasionally may appear as a solid, dark ring of material. Cytoplasm, usually coarsely granular and vacuolated, frequently contains bacteria, yeasts, and other debris.

Cysts. — Cysts, which have a wide size range of 10–35 μ, usually are 15–25 μ. They are most often spherical but may be oval-shaped or have other shapes. Mature cysts typically have eight nuclei, although supernucleate cysts may have 16 or more. The nuclei usually are visible in unstained cysts. In stained preparations, the nuclear characteristics are not as well defined as in trophozoites. Karyosomes may be compact or diffuse, and may or may not be eccentrically located. Peripheral chromatin may range from coarse and irregularly clumped granules to an appearance more uniform that that seen in trophozoites.

The cytoplasm of mature cysts may contain diffuse glycogen. In immature cysts, the glycogen may be a large, discrete mass, and the nuclei may be displaced to the sides of the cyst. Chromatoid bodies, which are seen less frequently in these cysts than in E. histolytica cysts, usually are splinter-like, with pointed ends.

LIFE CYCLE. — Direct transmission by ingestion of cyst stage.

DIAGNOSIS. — Demonstration of trophozoites or cysts in feces.

Diagnostic problems. — Some specimens may have nuclei containing central karyosomes and uniform, peripheral chromatin and may be confused with E. histolytica. Immature cysts containing four or fewer nuclei may be extremely difficult to distinguish from E. histolytica, as karyosomes in cyst nuclei frequently are centrally located and the distribution of peripheral chromatin may be uniform. Proper fixation and subsequent good staining of E. coli cysts is more difficult than with other amebic cysts.

FIGURES. — Plate 6:1–4; Plate 7:1–4; Plate 8:1–6.

PLATE 6

Entamoeba coli, Trophozoites, Trichrome Stain

Figs 1–3. These three trophozoites demonstrate some typical features of this parasite. The nuclei in all three have karyosomes that are eccentrically positioned, and peripheral chromatin that is irregularly distributed. Figure 2 best illustrates these features. Note that the cytoplasm is extensively vacuolated in all three trophozoites. Usually, the cytoplasm stains darker than that of E. histolytica.

Fig 4. This trophozoite has ingested a spherical cluster of minute spores of the fungus Sphaerita. Red blood cells are never seen in the cytoplasm of E. coli, however, bacteria, yeasts, and other microorganisms are commonly found. The nucleus in this trophozoite also demonstrates coarse, irregular, peripheral chromatin and a large, diffuse karyosome.

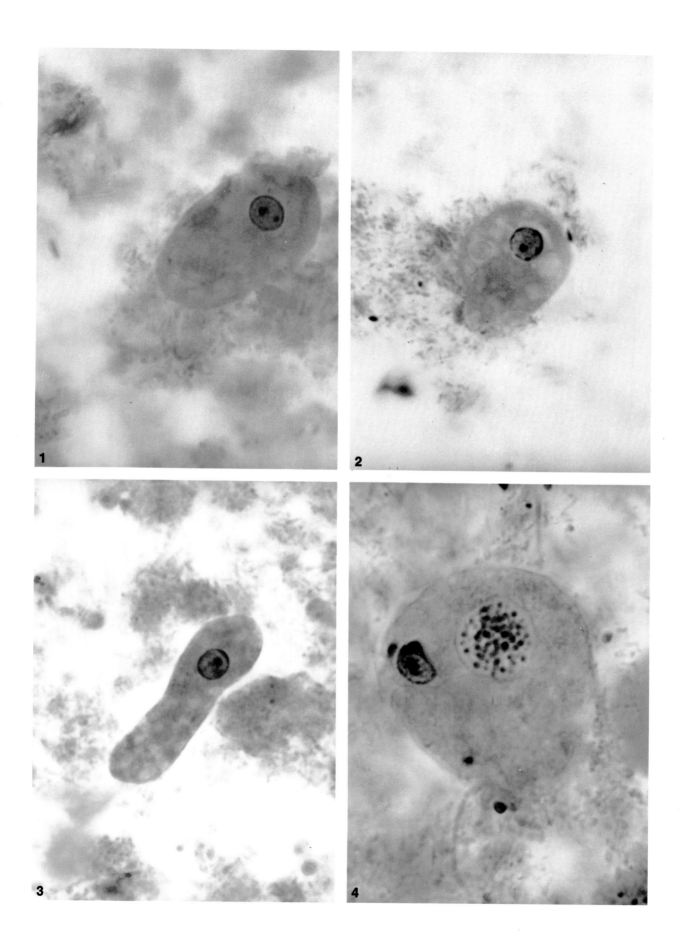

PLATE 7

Entamoeba coli, Cysts

Fig 1.   This typical, mature, iodine-stained cyst contains eight nuclei, five of which can be seen in this plane of focus. Note the smooth cyst wall and the refractile nature of the nuclei. The karyosomes in these nuclei appear to be centrally located, or nearly so. Because nuclear characteristics of *Entamoeba* cysts may vary, they may be unreliable for species identification, especially in wet mount preparations.

Fig. 2.   In this unstained, formalin-preserved cyst, the refractile nature of the cyst wall and of the nuclei is evident. Only four of eight nuclei are visible.

Fig 3.   The presence of mature cysts of *E. coli* and of *E. histolytica* in the same field emphasizes their differences in size and number of nuclei. Although these two species can be differentiated in this iodine-stained preparation, a permanent stained smear is necessary for confirmation.

Fig 4.   A cluster of iodine-stained cysts from a zinc sulfate flotation procedure shows the variation in size and shape that can be observed.

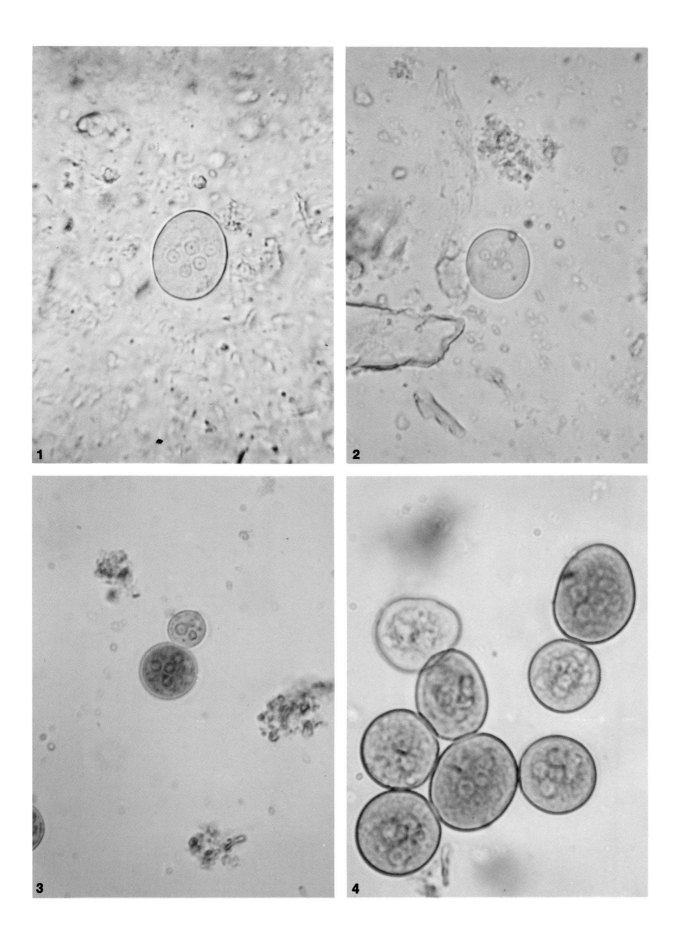

PLATE 8

Entamoeba coli, Trophozoites and Cysts, Trichrome Stain

Fig 1.   A uninucleate cyst in which a large, central, glycogen vacuole has compressed the nucleus to the margin.

Fig 2.   In this field one sees a trophozoite and a mature cyst, the latter with most of the eight nuclei and a chromatoid body visible. The nucleus of the trophozoite, which is typical of this species, shows the heavy, irregular, peripheral chromatin.

Fig 3.   A mature cyst in which five or six nuclei and a chromatoid body are visible.

Fig 4.   A trophozoite with a nucleus in which the karyosome appears large and irregular.

Fig 5.   Another mature cyst showing the characteristic chromatoid bodies with splintered ends.

Fig 6.   This rounded trophozoite, shown at somewhat lower magnification, illustrates *E. coli*'s tendency to stain more darkly than does *E. histolytica.*

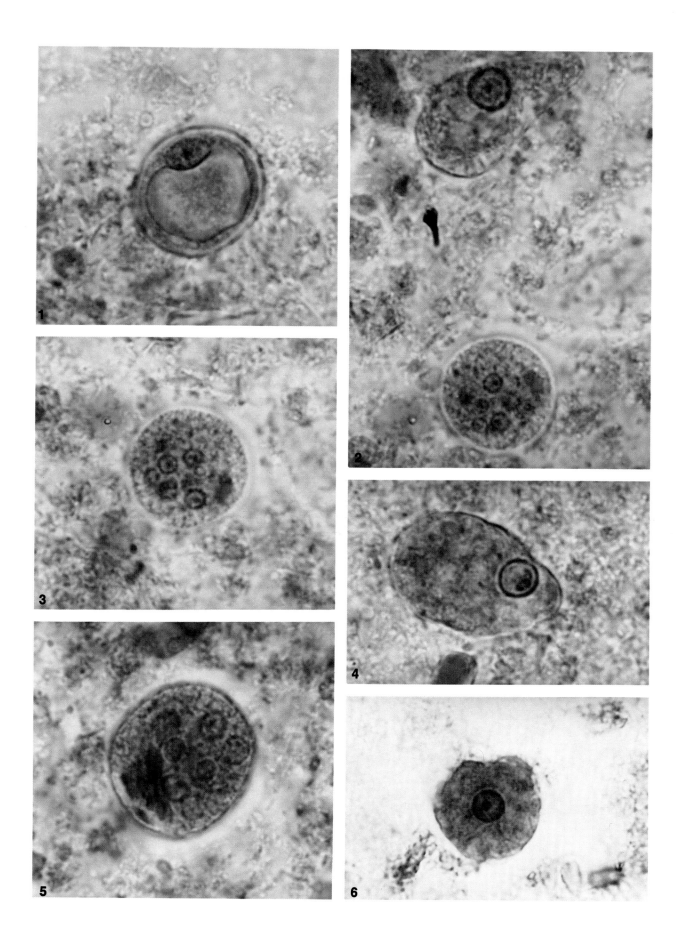

Endolimax nana

CLASSIFICATION. — Ameba.

DISEASE. — Nonpathogenic.

GEOGRAPHIC DISTRIBUTION. — Cosmopolitan.

LOCATION IN HOST. — Lumen of colon and cecum.

MORPHOLOGY. — Trophozoites. — This stage is small, measuring 6–12 μ, with an average range of 8–10 μ. Living trophozoites are sluggish and generally nonprogressive. The single nucleus sometimes is visible in unstained preparations. In stained organisms, the karyosome usually is large and irregularly shaped, but occasionally it may be fragmented or placed against one side of the nuclear membrane. There is no peripheral chromatin on the nuclear membrane. The cytoplasm, which is coarsely granular and often highly vacuolated, may contain bacteria.

Cysts. — Cysts are small, with a spherical to ellipsoidal shape. Mature cysts contain four nuclei; immature cysts are rarely seen. These cysts measure 5–10 μ, with a usual range of 6–8 μ. In stained preparations, the nucleus has a distinct karyosome that, while not as large as that seen in the trophozoite, is still larger than the karyosome of the *Entamoeba* species. Peripheral chromatin is absent. Although the nuclei are not visible in unstained preparations, the karyosomes are readily apparent in iodine-stained wet mounts. The cytoplasm may contain diffuse glycogen, and chroma-

toid bodies are absent. Occasionally, small granules or inclusions occur in the cytoplasm.

LIFE CYCLE. — Transmission by ingestion of cyst stage.

DIAGNOSIS. — Demonstration of trophozoite or cyst stages in feces.

Diagnostic problems. — Because of its small size, *E. nana* frequently is confused with other small amebae. Careful study of the nucleus of the trophozoite stage should readily separate it from *Entamoeba hartmanni* and *E. histolytica*. However, *E. nana* often is difficult to distinguish from *Iodamoeba*. In *E. nana*, the karyosome frequently is more irregularly shaped or applied against the nuclear membrane. When this feature is not present, these two species, based on a single organism, may be indistinguishable. Achromatic granules surrounding the karyosome, or between one side of the karyosome and the nuclear membrane, occur in *Iodamoeba* but are absent in *E. nana*.

There frequently is confusion in distinguishing cysts of *E. nana* from those of *E. hartmanni* and *E. histolytica*, however, the punctate karyosomes of the former are larger than those of *Entamoeba*. There should be no problems in distinguishing the cysts of *E. nana* and *Iodamoeba* since the mature cyst of *E. nana* has four nuclei while *Iodamoeba* has only one. Additionally, the cyst of *Iodamoeba* has the characteristic glycogen vacuole that is lacking in *E. nana*.

FIGURES. — Plate 9: 1–6; Plate 10: 1–6.

PLATE 9

Endolimax nana, Trophozoites and Cysts

Figs 1 and 2. Typical small trophozoites in trichrome-stained preparations. Note the vacuolated cytoplasm and nuclei with large, heavily-stained karyosomes. The nucleus is devoid of peripheral chromatin, as is characteristic for this species.

Fig 3. This trichrome-stained trophozoite has a nucleus in which the karyosome is band-like. Atypical nuclei are not uncommon in this species.

Fig 4. Three trophozoites are seen in this field. The central trophozoite has a nucleus with an atypical karyosome, whereas the organisms on either side of it have normal nuclei.

Fig 5. Three of the four nuclei are clearly visible in this mature cyst stained with iodine. Note the large karyosome and the lack of peripheral chromatin.

Fig 6. In this trichrome-stained cyst, all four nuclei are visible. The elongate shape of this cyst is not uncommon for the species.

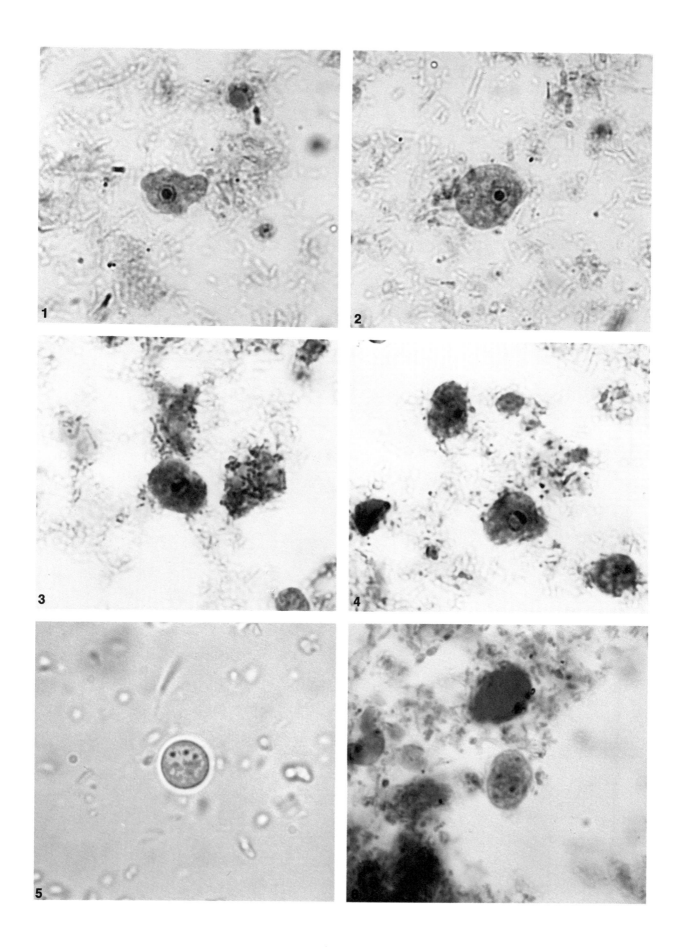

PLATE 10

Endolimax nana, Trophozoites and Cysts, Iron Hematoxylin Stain

Figs 1–4. All four of these trophozoites have the characteristic nuclear structure, ie, a large, heavily staining karyosome and a membrane lacking peripheral chromatin. The cytoplasm of each specimen also shows varying degrees of vacuolation.

Figs 5 and 6.   In these mature cysts, the typical nuclear structure is evident.

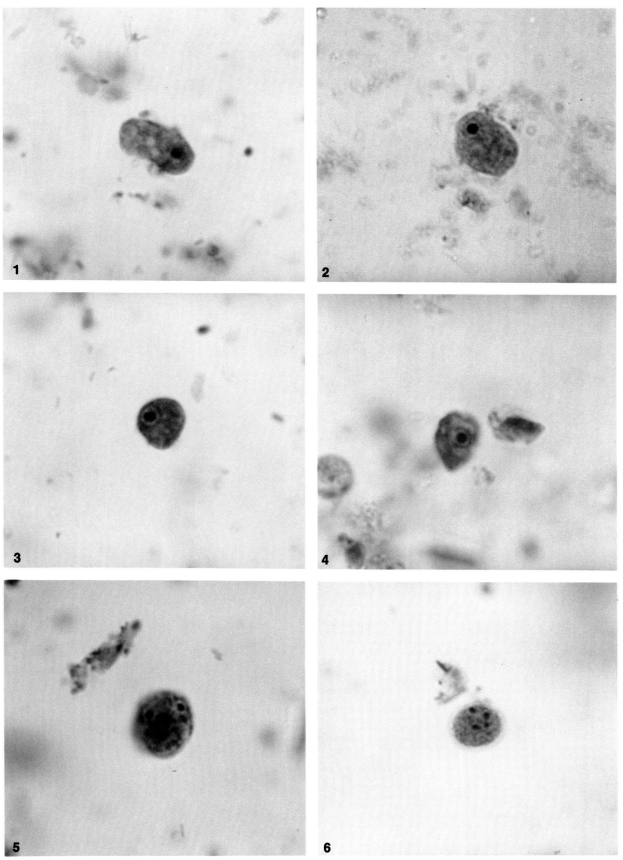

Iodamoeba bütschlii

CLASSIFICATION. — Ameba.

DISEASE. — Nonpathogenic.

GEOGRAPHIC DISTRIBUTION. — Cosmopolitan.

LOCATION IN HOST. — Lumen of colon and cecum.

MORPHOLOGY. — Trophozoites. — This stage measures 8–20 μ, with a mean measurement of 12–15 μ. They are sluggish and have nonprogressive movement. The single nucleus is not visible in unstained preparations. When stained, the karyosome of the nucleus is large and usually centrally located. Frequently, refractile achromatic granules, which are difficult to see, surround the karyosome or they lie between the margin of the karyosome and the nuclear membrane. No peripheral chromatin is on the nuclear membrane. The cytoplasm usually is coarsely granular, vacuolated, and may contain bacteria, yeasts, or other debris.

Cysts. — The shape of cysts varies considerably — from spherical to ellipsoidal. Their size may be 5–20 μ, with an average range of 10–12 μ. The mature cyst has only a single nucleus that is not visible in either unstained or iodine-stained preparations. When the organism is permanently stained, the nucleus is seen to contain a large, usually eccentric karyosome, and achromatic granules may or may not be around the karyosome or to one side of it. The most striking feature of the cyst is the presence of a discrete, compact mass of glycogen in the cytoplasm. In a wet specimen with iodine stain, the glycogen vacuole takes on a brownish-red color. The glycogen vacuole does not stain with permanent stains, but appears as a nonstaining, well defined mass.

LIFE CYCLE. — Transmission by ingestion of cyst stage.

DIAGNOSIS. — Demonstration of trophozoites or cysts in feces.

Diagnostic problems. — Most identification problems occur in the trophozoite stage. This species may be difficult and sometimes impossible to distinguish from Endolimax nana because their sizes may overlap, and both have large karyosomes. Although the presence of achromatic granules around the karyosome is helpful in identifying the organisms as Iodamoeba, these granules cannot always be seen. The cyst stage rarely poses a diagnostic problem since the glycogen mass stains so distinctly. However, one must remember that glycogen occasionally may occur as a discrete mass in other species of amebae.

FIGURES. — Plate 11:1–6; Plate 12:1–6.

PLATE 11

Iodamoeba bütschlii, Trophozoites and Cysts

Figs 1 and 3. These two trichrome-stained trophozoites demonstrate characteristic vacuolation of the cytoplasm and the typical nucleus containing a large, dark-staining karyosome. The nuclear membrane lacks peripheral chromatin. Trophozoites of Iodamoeba are difficult to differentiate from those of E. nana. Iodamoeba trophozoites may be larger, and their cytoplasm often contains abundant debris. The karyolymph space may be hazy and indistinct.

Figs 2 and 4. The most characteristic feature of Iodamoeba cysts is the large glycogen vacuole clearly seen in each of these two trichrome-stained organisms. These cysts are invariably uninucleate and have the typical large karyosome.

Figs 5 and 6. Cysts stained in iodine, shown here at low and high magnification, are readily identified by the dark-staining glycogen mass. Nuclei in these cysts are not easily seen with iodine stain. The cysts' affinity for iodine has led to their being referred to as "iodine cysts."

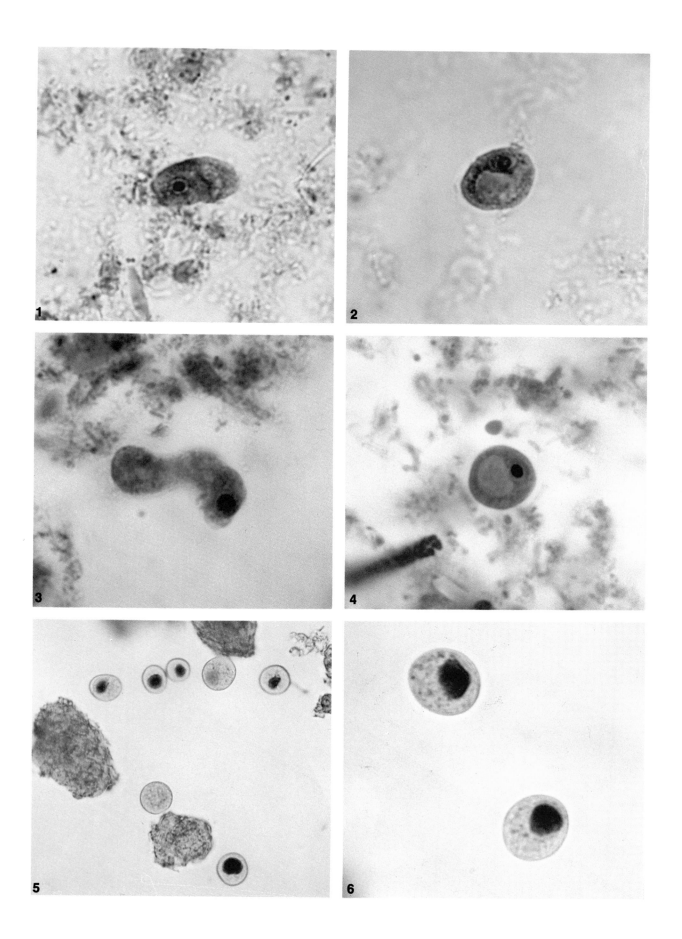

PLATE 12

Iodamoeba bütschlii, Trophozoites and Cysts, Iron Hematoxylin Stain

Figs 1, 3, and 5. Trophozoites shown here display the characteristic morphologic features of this species. Note the vacuolated cytoplasm and the nucleus with a large, darkly stained karyosome in each trophozoite.

Figs 2, 4, and 6. Typical uninucleate cysts. Note the nuclear features and the large glycogen vacuoles.

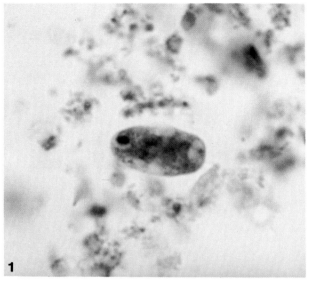

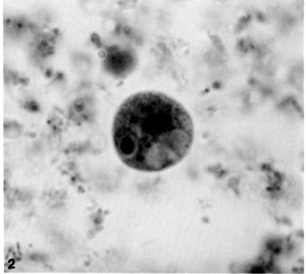

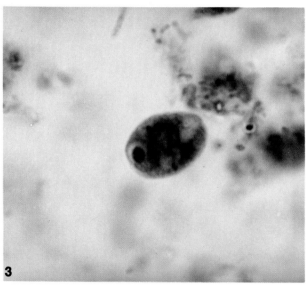

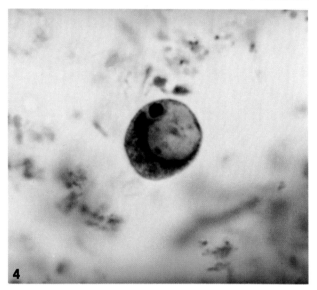

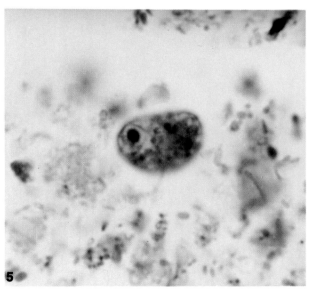

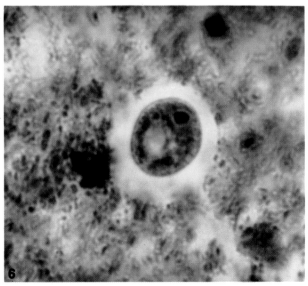

Naegleria fowleri

CLASSIFICATION. — Ameba.

DISEASE. — Primary amebic meningoencephalitis.

GEOGRAPHIC DISTRIBUTION. — Cosmopolitan.

LOCATION IN HOST. — Central nervous system, spinal fluid, and perhaps other organs.

MORPHOLOGY. — Trophozoites. — Ameboid and ameboflagellate trophozoites are free-living in nature: in ponds, lakes, and temporary pools of fresh water. Only ameboid trophozoites, which usually range from 10 to 20 μ, occur in the human host. Their movement in wet mounts of spinal fluid may be rapid and directional. The cytoplasm is granular and the nucleus is relatively large, containing a large central karyosome. The nuclear membrane lacks peripheral chromatin.

Cysts. — Although cysts occur in the external environment, Naegleria does not form cysts in human tissue.

LIFE CYCLE. — The entire life cycle takes place in the external environment. Apparently, humans usually acquire this infection by swimming in very warm bodies of fresh water, whereupon the amebae invade the olfactory mucosa and then enter the brain. Acute, fulminating infection generally occurs, and the host usually dies within one to two weeks.

DIAGNOSIS. — Antemortem diagnosis of this disease often is difficult. Primary amebic meningoencephalitis is suspected in those cases in which patients had recently been swimming in warm water. Failure to demonstrate bacteria in purulent cerebrospinal fluid and the finding of amebae with nuclei containing large karyosomes are indicative of this parasite. Usually, the amebae with characteristic nuclei are readily found in brain tissue sections from fatal cases.

Diagnostic problems. — Frequently, the course of the infection is so rapid that death ensues before this particular diagnosis is considered. Detecting amebae in cerebrospinal fluid often is difficult.

COMMENTS. — Other free-living amebae (Acanthamoeba, Hartmannella) may cause human infection. Whereas Naegleria produces acute, fulminating infection, Acanthamoeba may produce chronic infections that may or may not be fatal. In Acanthamoeba infections, both trophozoite and cyst stages may occur in brain tissue. Acanthamoeba usually is larger (25 – 35 μ) than Naegleria, and the cyst has a double wall.

FIGURES. — Plate 13:1 – 4.

PLATE 13

Naegleria fowleri, Trophozoites

Fig 1. Trophozoite stage, trichrome-stained, in culture medium. This large, free-living ameba has a very large nucleus with a large, dark-staining karyosome, and no peripheral chromatin. Note the pseudopods of this ameba.

Fig 2. As the trophozoite grows and prepares to divide in culture, two nuclei may be seen. Trophozoites from culture are larger than those usually observed in humans.

Fig 3. This trophozoite from human spinal fluid shows the typical nucleus with a large karyosome. A leukocyte lies adjacent to the ameba, indicating the relative size of the two. Organisms seen in human spinal fluid and tissues are typically smaller than those seen in culture.

Fig 4. A trophozoite in human brain tissue from a fatal case of primary amebic meningoencephalitis. Note the trophozoite, readily identified by its large nucleus and prominent karyosome, in the center of the field.

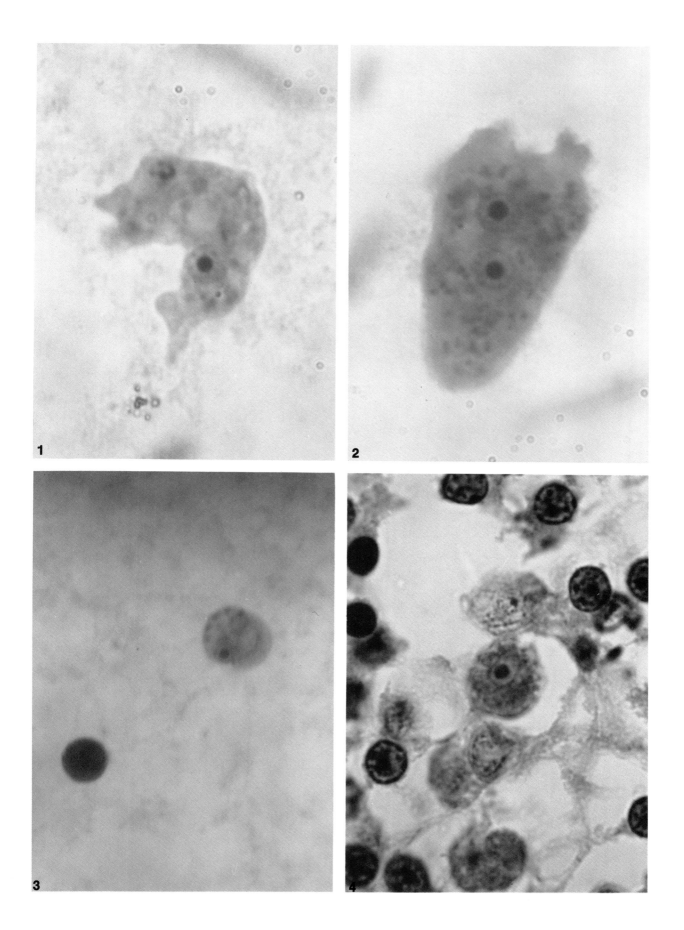

Dientamoeba fragilis

CLASSIFICATION. — Flagellate.

DISEASE. — Although many workers consider this species to be nonpathogenic, it apparently may cause diarrhea and abdominal discomfort in some people.

GEOGRAPHIC DISTRIBUTION. — Cosmopolitan.

LOCATION IN HOST. — Colon.

MORPHOLOGY. — Trophozoites. — The trophozoite stage, similar to that of an ameba, has no evident flagella. Pseudopodia are angular or broad-lobed and transparent. Motility generally is non-progressive. Organisms are 5–15 μ, with a usual range of 9–12 μ. Most organisms have two nuclei, although approximately 30%–40% are uninucleate. Nuclei are not visible in unstained preparations. In stained organisms, the nuclei have a central mass of karyosomal material that usually is in a cluster of four to eight granules. There is no peripheral chromatin on the nuclear membrane. Cytoplasm may be finely to heavily granular and may be vacuolated; it may contain bacteria, yeasts, and other debris.

Cysts. — This species has no cyst stage.

LIFE CYCLE. — Direct transmission, probably by ingestion of trophozoite stage. However, some workers have suggested that the organism might be transmitted within the eggs of some helminths, especially pinworm eggs.

DIAGNOSIS. — Demonstration of trophozoite in feces.

Diagnostic problems. — Organisms frequently are pale-staining and can be easily overlooked. Uninucleate trophozoites pose more identification problems than do the binucleate forms.

FIGURES. — Plate 14:1–6.

PLATE 14

Dientamoeba fragilis, Trophozoites, Trichrome Stain

Figs 1, 2, and 4. These three organisms, each with two nuclei, are typical of this species. In at least one nucleus in each organism, the karyosome is fragmented into four segments, as is characteristic for this species.

Figs 3, 5, and 6. These three organisms each have a single nucleus. Note that all the nuclei have segmented karyosomes (best illustrated in Fig 6). In general, about 60% of *Dientamoeba* organisms will have two nuclei. *Dientamoeba* tends to take a rather pale stain and, consequently, it often can be overlooked in stained fecal smears.

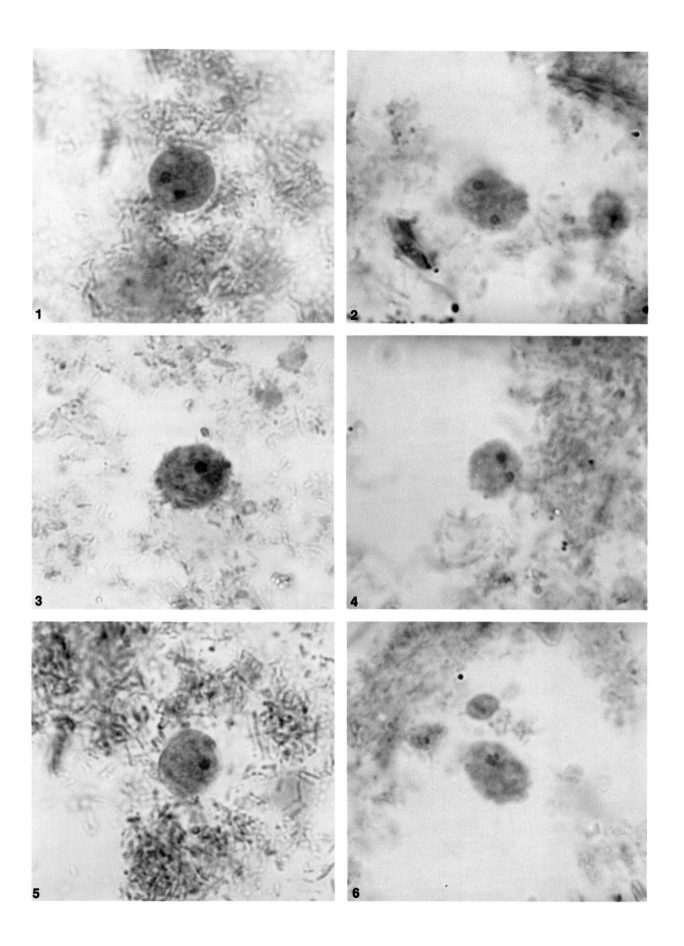

Giardia lamblia

CLASSIFICATION. — Flagellate.

DISEASE. — Giardiasis.

GEOGRAPHIC DISTRIBUTION. — Cosmopolitan.

LOCATION IN HOST. — Small intestine.

MORPHOLOGY. — Trophozoites. — Trophozoites are pear-shaped organisms, measuring 10–20 μ, with the usual range being 12–15 μ. When seen free from debris, the tumbling kind of motility of the trophozoite stage is likened to that of a falling leaf. This bilaterally symmetrical organism has two nuclei that are not visible in unstained or iodine-stained wet mounts. Just posterior to the nuclei are a pair of sausage-shaped bodies lying transversely in the cytoplasm. A concavity or bowl-shaped depression — referred to as the "sucking disk" — occupies the ventral surface of the anterior part of the body. This disk serves as the organism's site of attachment to the mucosal epithelium.

The eight flagella — four of which are lateral, two ventral, and two caudal — are continuations of the axonemes. Each arises from a blepharoplast. In living trophozoites, the movement of the flagella is visible, however, the flagella are not seen in stained preparations unless special stains are used. Stained organisms show two nuclei, one on each side of the midline, and these have central karyosomes with no peripheral chromatin.

Cysts. — Cysts are ovoid to ellipsoid in shape and measure 8–19 μ, the usual range being 11–14 μ. Mature cysts have four nuclei, whereas the immature organisms have two. Nuclei and intracytoplasmic fibrils are visible in iodine-stained wet mounts. In stained preparations, nuclei are concentrated toward the broader part of the cyst; fibrils that may cross one another are located more toward the posterior end. The cytoplasm of the cyst may retract from the cyst wall, especially in formalin-preserved specimens.

LIFE CYCLE. — Direct transmission by ingestion of cyst stage. Animal reservoirs may be of considerable significance in human infection.

DIAGNOSIS. — By demonstration of trophozoites and cysts in feces. Duodenal aspirates and duodenal "capsule" technique (Entero-Test) also may be used to detect organisms.

Diagnostic problems. — *Giardia* rarely poses diagnostic difficulties when either the trophozoite or cyst stage is found. Most difficulties stem from the fact that organisms may be difficult to detect in some clinical cases, and one must resort to examining multiple specimens or to doing duodenal aspirates.

FIGURES. — Plate 15:1–6; Plate 16:1–6.

PLATE 15

Giardia lamblia, Trophozoites and Cysts, Trichrome Stain

Fig 1.   The trophozoite typically has a pyriform shape and two anteriorly placed nuclei, one on either side of the longitudinally oriented axonemes. These nuclei have two centrally located karyosomes. Flagella are not seen because they stain poorly.

Fig 2.   In this organism, the karyosomes are fragmented in the two nuclei. The dark-staining, parabasal body posterior to the nuclei gives the organism the characteristic appearance of a smiling face.

Figs 3–6.   Note the structure of *Giardia* cysts, with varying degrees of clarity in each illustration. Mature cysts typically have four nuclei, paired axonemes, and fibrils. The halo effect around the organisms, seen in each of the figures, is a result of shrinkage due to fixation.

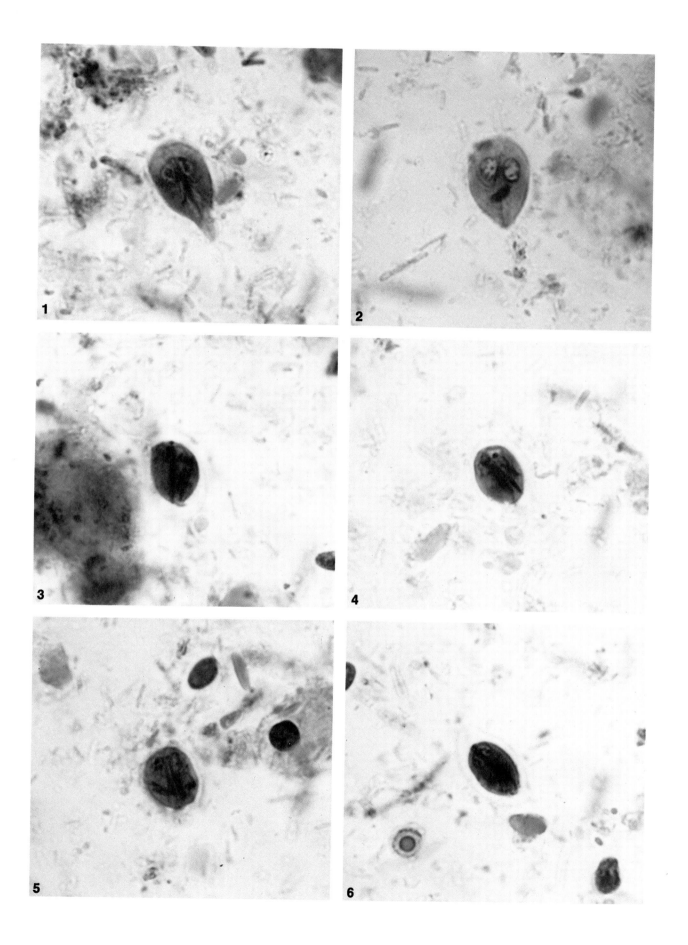

PLATE 16

Giardia lamblia, Trophozoites and Cysts, Iron Hematoxylin Stain

Figs 1–3. These three trophozoites illustrate characteristic features of this species. Frequently, only the anterior portion of the trophozoite is evident (Fig 1). The organism in Figure 3 is seen in its lateral aspect, with the ventral portion representing the "sucking disk," with which the parasite adheres to the mucosal epithelium of the intestine.

Figs 4–6. These cysts have the characteristic features of the cyst stage. The disparity in size between Figures 5 and 6, as compared to Figure 4, is due to shrinkage of the organisms during fixation, as these all were photographed at the same magnification.

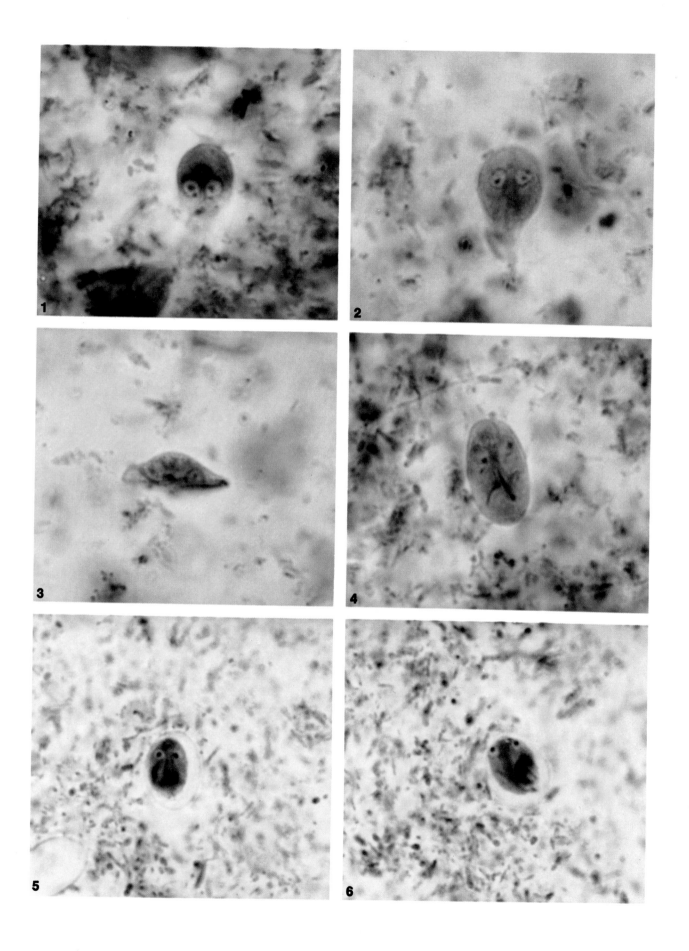

Chilomastix mesnili

CLASSIFICATION. — Flagellate.

DISEASE. — Nonpathogenic.

GEOGRAPHIC DISTRIBUTION. — Cosmopolitan.

LOCATION IN HOST. — Primarily in large intestine but may occur in small intestine.

MORPHOLOGY. — Trophozoites. — These pear-shaped organisms are 6–24 μ long, with a usual range of 10–15 μ. Living trophozoites have a stiff, rotary movement, and the single nucleus is not visible in unstained preparations. In stained organisms, a prominent cytostome may be seen extending ⅓–½ of the length of the body with a spiral groove extending across the ventral surface. The nucleus has a large karyosome situated centrally or against the nuclear membrane. Peripheral chromatin is generally evenly distributed.

Cysts. — The uninucleate cyst is lemon-shaped, with an anterior hyaline knob. Cysts average 7–9 μ but may range from 6 to 10 μ. Fibrils in the cyst frequently give the appearance of an open safety pin alongside the cytostome. Peripheral chromatin may be concentrated to one side of the nucleus.

LIFE CYCLE. — Direct transmission by ingestion of the cyst stage.

DIAGNOSIS. — Demonstration of trophozoites and cysts in feces.

Diagnostic problems. — Trophozoites and cysts may take a pale stain and may be easily overlooked. In stained preparations, *Chilomastix* sometimes is confused with *Entamoeba histolytica* and *E. hartmanni*, however, the presence of the cytostome and the characteristics of the nucleus should allow for proper identification.

FIGURES. — Plate 17:1–6; Plate 18:1–6.

PLATE 17

Chilomastix mesnili, Trophozoites and Cysts

Figs 1 and 2. Note the enlongated, pyriform shape of the trophozoites in these trichrome preparations. The pointed, posterior end is clearly evident in Figure 1. The nucleus is at the anterior end, and the saclike cytostome can be seen in both organisms. The flagella cannot be seen, due to their poor staining qualities.

Figs 3–5. The cyst stage of this trichrome-stained parasite (Figs 3 and 4) usually is lemon-shaped with a lightly stained nipple-like prominence at one end. Spherical organisms are those likely to be misidentified as amebae, and due to its orientation, the cyst frequently may appear spherical, as in the iron hematoxylin-stained organism illustrated in Figure 5. The single nucleus and fibrils are clearly seen in all of these organisms.

Fig 6. This iodine-stained cyst shows all the typical features of the parasite.

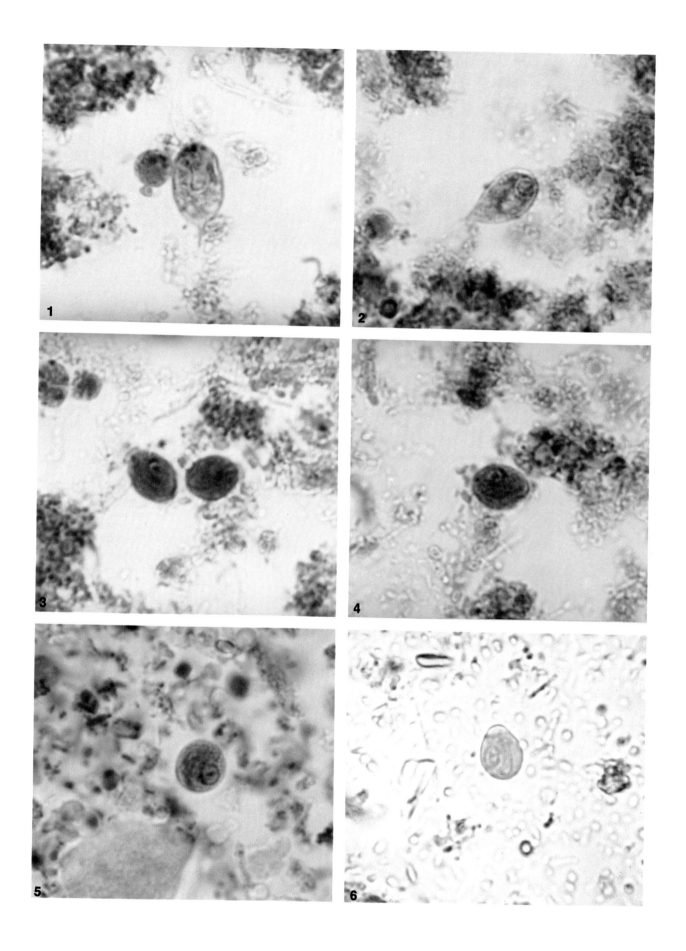

PLATE 18

Chilomastix mesnili, Trophozoites and Cysts, Iron Hematoxylin Stain

Figs 1–3.   These trophozoites all show the characteristic form of the parasite. Note the anteriorly placed nucleus and the pointed, posterior end. The cytostome is poorly demarcated in these particular organisms.

Figs 4–6.   In these examples, both lemon-shaped and spherical cysts display the same morphologic features described in the trichrome-stained specimens in Plate 17. The nuclei and fibrils are clearly seen.

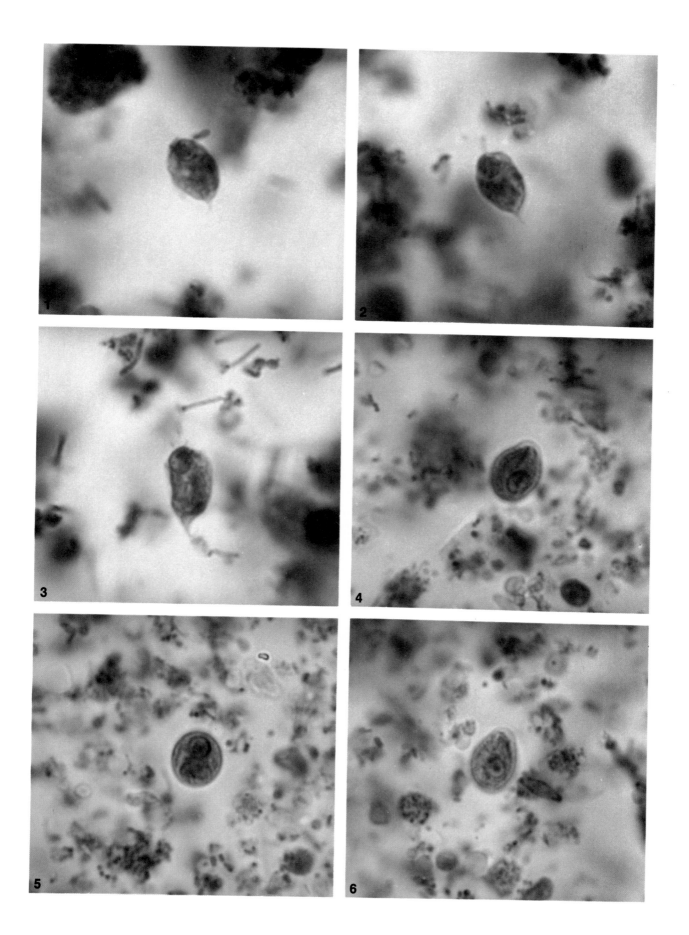

Balantidium coli

CLASSIFICATION. — Ciliate.

DISEASE. — Balantidiasis.

GEOGRAPHIC DISTRIBUTION. — Wide distribution in temperate and warm climates. In tropical areas, nonhuman primates commonly are infected, whereas pigs are infected in other areas of the world. However, neither monkeys nor pigs appear to be significant reservoirs of human infection.

LOCATION IN HOST. — Colon.

MORPHOLOGY. — Trophozoites. — The trophozoite is a large, ovoid, ciliate parasite measuring $50-200\ \mu$, the average being $50-100\ \mu$, with widths of $40-70\ \mu$. Living trophozoites have a rotary, boring motion, and they may move very rapidly across the field of view. In living specimens, the cilia maintain a constant, synchronized motion. The trophozoite is somewhat pointed at the anterior end, with the cytostome — a deep, somewhat curved depression — located toward the anterior end. Two nuclei are present — a kidney bean-shaped macronucleus that often is visible in unstained preparations, and a smaller micronucleus that is difficult to discern even in stained organisms. The cytoplasm may contain numerous food and contractile vacuoles. Contractile vacuoles empty through the cytopyge, a small opening at the posterior end.

Cysts. — Cysts are spherical to oval-shaped, usually measuring $50-70\ \mu$. Cilia often are visible through a tough cyst wall. Since nuclear multiplication does not occur in the cyst stage, both the large macronucleus and small micronucleus are present. Cytoplasmic inclusions are not extruded during encystment so that both food and contractile vacuoles can be seen in younger cysts, whereas older cysts may have a granular appearance.

LIFE CYCLE. — Direct transmission by ingestion of cyst stage.

DIAGNOSIS. — Demonstration of trophozoites and cysts in feces.

Diagnostic problems. — The size and characteristic morphology of Balantidium essentially preclude any diagnostic difficulties.

FIGURES. — Plate 19:1-4.

PLATE 19

Balantidium coli, Trophozoites and Cyst

Figs 1 and 3. These ciliate organisms in the trophozoite stage are readily recognized by their large size, ciliate surface, and prominent cytostome — even in the unstained, saline preparations shown here. Portions of the nuclei are visible at the posterior end of the organisms.

Fig 2. The Balantidium cyst consists of a clear, refractile wall enveloping the ciliate. These cysts are spherical, as is the organism within. In some instances, ciliary structures can be seen on the surface of the organism within the cyst (not on the outside of the cyst wall). In this illustration, however, these structures are not clearly seen. The cytoplasm is markedly granular. The macronucleus is represented by the large, less granular area in the center of this organism.

Fig 4. This trichrome-stained, ciliate trophozoite illustrates the key morphologic features of the organism. Note the prominent cytostome and the large, red-stained macronucleus. The micronucleus is not visible.

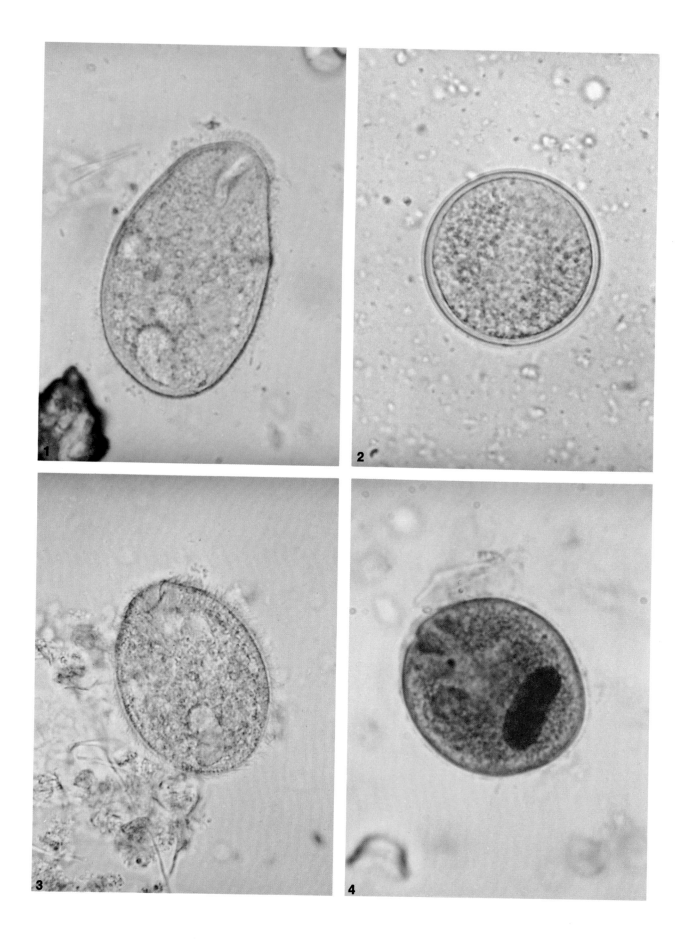

Isospora and Sarcocystis

CLASSIFICATION. — Coccidia.

DISEASES. — Coccidiosis; *Sarcocystis* infection.

GEOGRAPHIC DISTRIBUTION. — Cosmopolitan.

LOCATION IN HOST. — *Isospora belli* and one or more species of *Sarcocystis* occur in the human intestinal tract. Other species of *Sarcocystis* may occur in human musculature.

MORPHOLOGY. — Trophozoites. — These stages are rarely, if ever, observed in *Isospora* or *Sarcocystis* infections of the intestine.

Cysts. — In the oocyst stage, *I. belli* measures 20–33 μ long by 10–19 μ wide, and is unsporulated when passed in feces. The oocysts are ovoid, taper at the ends, and have a smooth, double-layered hyaline wall.

In those species of *Sarcocystis* producing human intestinal infections, ie, *S. bovihominis, S. suihominis,* and others, oocysts and/or sporocysts may be found in feces. Oocysts are thin-walled and contain two sporocysts, each of which has four sausage-shaped sporozoites and a refractile residual body. The two sporocysts lie side by side within the oocyst so that the oocyst measures 15–19 μ long by 15–20 μ wide. Individual sporocysts measure 15–19 μ long by 8–10 μ wide.

LIFE CYCLE. — *I. belli* is the only species of the genus reported to occur in humans. The life cycle is direct, with no intermediate host involvement. Unsporulated oocysts undergo development in the soil, contain two sporulated sporocysts in the infective stage. Sexual reproduction takes place in the intestinal mucosa, and oocysts are produced and passed in feces.

All species of *Sarcocystis* have an obligatory two-host life cycle. The intermediate host acquires infection by ingestion of sporulated sporocysts, and tissue cysts (sarcocysts) that eventually are produced will infect the second or definitive host when ingested. Ingestion of these zooite-containing tissue cysts by the definitive host results in sexual reproduction in the intestinal epithelium to produce oocysts that are then passed in feces. Humans may serve as the definitive host for several species of *Sarcocystis* and may also serve as a dead-end, intermediate host for other species of the genus.

DIAGNOSIS. — For those species in which humans serve as the definitive host, diagnosis is based on demonstration of oocysts and/or sporocysts in feces. When man serves as the intermediate host for species of *Sarcocystis*, diagnosis is made by finding tissue cysts in histologic sections of muscle.

Diagnostic problems. — Oocysts and/or sporocysts may be overlooked in fecal examinations. The passage of these stages may be short and sporadic so that multiple stool examinations may be required. Tissue cysts usually are an incidental finding in histologic specimens of muscle.

FIGURES. — Plate 20:1–4.

PLATE 20

Isospora belli and *Sarcocystis bovicanis,* Oocysts and Sporocysts

Fig 1. *Isospora belli,* immature oocyst. This small, formalin-fixed oocyst has a thin wall and contains a granular sporoblast. This is the diagnostic stage seen in freshly passed stool specimens.

Figs 2–4. *Sarcocystis bovicanis,* formalin-fixed oocysts and sporocysts. Although this coccidian is a parasite of dogs, the size and structure of the oocyst and sporocyst are similar — if not identical — to species occurring in the human intestine. The oocyst in freshly passed feces has undergone sporulation and contains two sporocysts. The thin-walled oocyst usually ruptures so that individual sporocysts are more commonly seen in feces. In Figure 2, note the presence of both a thin-walled oocyst *(arrow)* and two individual sporocysts containing sporozoites. Figures 3 and 4, at respective low- and high-power magnifications, show individual sporocysts, each containing four sporozoites and a residual body.

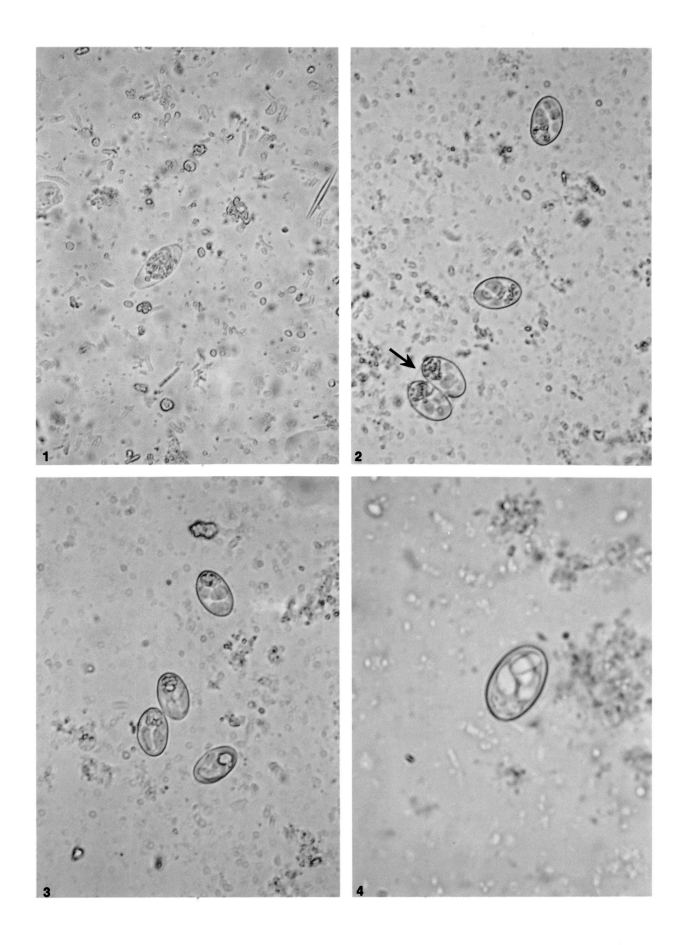

Toxoplasma gondii

CLASSIFICATION. — Coccidia.

DISEASE. — Toxoplasmosis.

GEOGRAPHIC DISTRIBUTION. — Cosmopolitan.

LOCATION IN HOST. — Trophozoite stages are found in various cells, tissues, and fluids. Cyst stages occur in the central nervous system, as well as in skeletal and cardiac muscles, and other visceral organs.

MORPHOLOGY. — Trophozoites. — These ovoid organisms have a pointed anterior end, a more blunt posterior end, and a large nucleus. They measure $4-8\ \mu$ long by $2-3\ \mu$ wide.

Cysts. — The cyst stage, that usually seen in humans, varies in size. It usually is spherical but may be elongate in cardiac and skeletal muscles. The cyst contains many zooites, referred to as bradyzoites.

LIFE CYCLE. — This coccidian parasite has a complex life cycle, felines being the only hosts in which sexual reproduction occurs in the intestinal epithelium. As a result, typical coccidian oocysts are passed in feces. Humans acquire infection by eating material contaminated by infected cat feces or by eating cysts in meat from chronically infected animals. Transplacental transmission is an important source of human infection.

DIAGNOSIS. — Since oocysts of *Toxoplasma* are found only in cat feces, the diagnosis of human infection is based mainly on symptoms and use of serologic procedures.

Diagnostic problems. — The trophozoite stages (referred to as tachyzoites) that are sometimes seen in exudates and tissue impression smears must be differentiated from *Leishmania, Histoplasma,* and *Pneumocystis.* Tissue cysts must be differentiated from amastigote nests of *Trypanosoma cruzi* and from cysts of *Sarcocystis.*

FIGURES. — Plate 21:1–4.

PLATE 21

Toxoplasma gondii, Trophozoites and Cysts

Fig 1. Cyst in hematoxylin-eosin-stained tissue section of mouse brain. This cyst form, characteristic of chronic infection, is found most frequently in the central nervous system, skeletal muscle, and heart. These cysts may contain hundreds of organisms called zooites. Cysts, which may persist for the life of the host, are capable of initiating active infection when the host's resistance is sufficiently reduced. Little if any inflammatory response occurs around intact cysts.

Fig 2. Many zooites can be seen in this Giemsa-stained cyst in an impression smear.

Fig 3. Giemsa-stained trophozoites in peritoneal fluid of mouse. These trophozoites — sometimes called tachyzoites — are found in a variety of tissues and fluids, usually during the acute stage of infection. Division normally occurs intracellularly, but rupture of the host cell may release organisms into the body fluids. In tissue sections, the organisms are more often spherical or elongate than crescent-shaped, as in tissue impression smears or films of body fluids. These forms can be distinguished from the amastigote stage of hemoflagellates by the absence of a kinetoplast.

Fig 4. Trophozoites in tissue culture. These Giemsa-stained trophozoites in tissue culture assume the typical shape described in Figure 3.

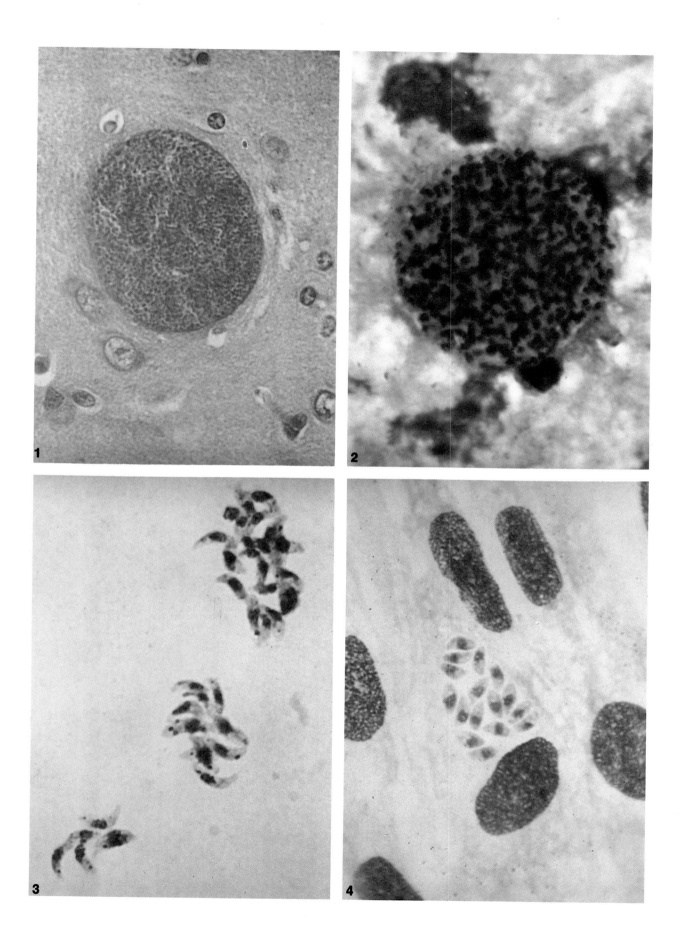

PLATE 22

Artifacts in Stained Fecal Smears

Fig 1. Note the two trichrome-stained epithelial cells in this field. Although the elongate cell should not cause diagnostic problems, the mucosal cell must be differentiated from a protozoan. The nucleus is large in relation to the amount of cytoplasm. Chromatin distribution does not have a pattern typical of any intestinal protozoan.

Fig 2. Six trichrome-stained yeasts are shown in this field. Note the variation in staining, with colors ranging from red to green. These yeasts should not be confused with protozoan cysts, however, their uniform structure and small size may pose identification problems for inexperienced microscopists.

Fig 3. Polymorphonuclear leukocytes (pus cells) in trichrome-stained fecal smear. The presence of massive numbers of polymorphonuclear leukocytes is less apt to confuse the viewer than would the finding of an isolated pus cell or two. Observe the relatively large, ringed nuclei and ragged cytoplasm. These cells have an approximate size of 14 μ.

Fig 4. These trichrome-stained organisms of *Blastocystis hominis* appear cyst-like, not only because they usually are spherical or nearly so, but also because they are refractile in unstained wet-mount preparations. They have a central vacuole, with large granules arranged around the periphery. When stained, as in this figure, the peripheral granules are red, and neither a clear space around the granules nor a nuclear membrane is visible. These organisms are of uncertain classification (but usually considered to be a yeast), and are often confused with amebic cysts. When exposed to tap water, living *Blastocystis* organisms rupture. Although within a given stool specimen *B. hominis* displays considerable size variation, it is within the size range for amebic cysts.

Fig 5. In this illustration of a macrophage and polymorphonuclear leukocytes stained with trichrome, the large macrophage in the center has a single nucleus and an ingested erythrocyte. Numerous polymorphonuclear leukocytes and a few erythrocytes also are present in this microscopic field.

Fig 6. In this field, two trichrome-stained *B. hominis* demonstrate the same characteristics described in Figure 4.

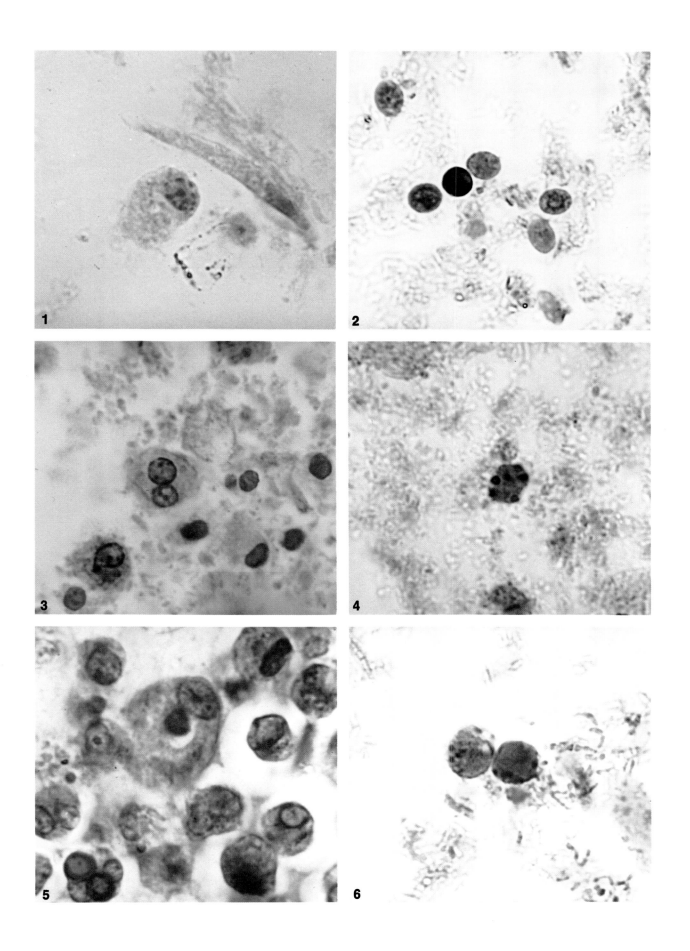

Trypanosoma gambiense, T. rhodesiense, and T. cruzi

CLASSIFICATION. — Flagellates.

DISEASES. — African trypanosomiasis or sleeping sickness (*T. gambiense* and *T. rhodesiense);* American trypanosomiasis or Chagas' disease *(T. cruzi).*

GEOGRAPHIC DISTRIBUTION. — African trypanosomiasis (*T. gambiense* and *T. rhodesiense*) is endemic in a belt across central Africa south of the Sahara from west Africa (*T. gambiense*) to east Africa (*T. rhodesiense*). *T. cruzi* occurs only in the western hemisphere — from the southern United States southward to Argentina.

LOCATION IN HOST. — Trypomastigote stages occur in the bloodstream in all species. In *T. cruzi* only, amastigote stages occur in reticuloendothelial tissues, glial cells of the central nervous system, and in cardiac muscle.

MORPHOLOGY. — Trypomastigotes. — The organism of this stage — often called the trypanosome — may range in shape from a delicate and spindle-shaped organism to one that appears broad and stumpy. The African species are indistinguishable from each other morphologically. Both measure 14–33 μ long, have a central nucleus, a small kinetoplast at the blunt posterior end, and a free flagellum coursing forward to leave the body at the anterior end. Dividing forms are found in the bloodstream.

T. cruzi is approximately 20 μ long. This organism has a central nucleus, a very large kinetoplast at the pointed posterior end, and a flagellum that courses forward to leave the body at the anterior end. Dividing forms do not occur in the bloodstream.

Other stages. — The trypomastigote stage is the only form occurring in the human host in the African trypanosomes. In addition to the trypomastigote stage in blood, *T. cruzi* has an amastigote stage that multiplies in reticuloendothelial tissues, glial cells, and cardiac muscle. These amastigotes occur as "nests" of cells. Promastigote and epimastigote stages occur in the insect vectors of both the African and American trypanosomes.

LIFE CYCLE. — African trypanosomes are transmitted by the bite of infected tsetse flies (*Glossina* spp). When flies ingest trypomastigotes and become infected, the flagellates multiply as epimastigote stages before migrating to the salivary glands to be inoculated as trypomastigotes.

Triatomid bugs are the intermediate hosts for *T. cruzi*. Bugs are infected when they feed on infected animals or humans, and ingest trypomastigote stages in blood or amastigote stages in macrophages or other cells. In the triatomid bug's midgut, these stages transform into epimastigotes that multiply and then migrate into the hindgut, transforming into infective trypomastigotes. These infective organisms are passed in the bug's feces, which are deposited as the bug feeds on the human or animal host. The deposited feces then may be rubbed into the bite wound or a skin abrasion. In the mammalian host, these organisms enter macrophages immediately, becoming amastigotes and beginning multiplication.

DIAGNOSIS. — African trypanosomiasis is diagnosed by finding trypomastigotes in blood during febrile periods, in lymph node aspirates in early stages of infection, or in cerebrospinal fluid. Serum IgM level determinations and fluorescent antibody tests are useful serologic diagnostic procedures. The finding of *T. cruzi* trypomastigotes in blood in early acute infections, or during febrile periods in chronic infections, is diagnostic of this infection. Blood cultures, animal inoculations, and serologic procedures also are useful diagnostic tools.

Diagnostic problems. — Trypomastigotes in both African and American trypanosomiasis may be sparse or absent in stained blood films. In South America, trypomastigotes of *T. cruzi* must be differentiated from a more rare human trypanosome infection, *T. rangeli* (not illustrated in this atlas). The latter is larger, has a free flagellum that is shorter than seen in *T. cruzi*, and the kinetoplast is smaller than that of *T. cruzi*.

FIGURES. — Plate 23:1–5; also see Plate 24:1.

PLATE 23

Trypanosoma gambiense and *T. cruzi,* Trypomastigotes and Epimastigotes

Fig 1. This is a thin blood film containing the trypomastigote stage of *T. gambiense.* These trypomastigotes, indistinguishable from those of *T. rhodesiense,* have a small kinetoplast near the somewhat blunted, posterior end, and a conspicuous, undulating membrane with a flagellum. At the extreme right margin note the two dividing forms; this is characteristic for the African trypanosomes, but is not seen in *T. cruzi.*

Fig 2. "Broad" form of *T. cruzi* trypomastigote in mouse thin blood film. Note the large kinetoplast close to the short, pointed, posterior tip of this organism. Broad forms, such as this, usually assume an S or C shape in stained preparations. The centrally located nucleus is subspherical, and the undulating membrane inconspicuous. *T. cruzi* does not undergo division in the trypomastigote stage.

Fig 3. "Slender" form of *T. cruzi* trypomastigote in mouse thin blood film. Slender forms are commonly found during acute infections. Note the elongate nucleus and the large kinetoplast followed by a long, pointed, posterior tip. The undulating membrane may be inconspicuous in stained preparations.

Fig 4. Two typical C forms are visible in this mouse thin blood film of trypomastigotes of *T. cruzi.* Note the very large kinetoplasts that seem to bulge from the body at the posterior end. In contrast, the kinetoplasts in the African trypanosomes are small and subterminal.

Fig 5. In this illustration of *T. cruzi* epimastigotes in culture, note the small, S-shaped trypomastigote above and to the left of center. It has a large kinetoplast near the posterior end and a flagellum that extends along its length, projecting beyond the anterior end. An undulating membrane is present between the flagellum and the body of the organism. This form—the infective stage for humans—is found in cultures and in the feces of infected triatomid bugs.

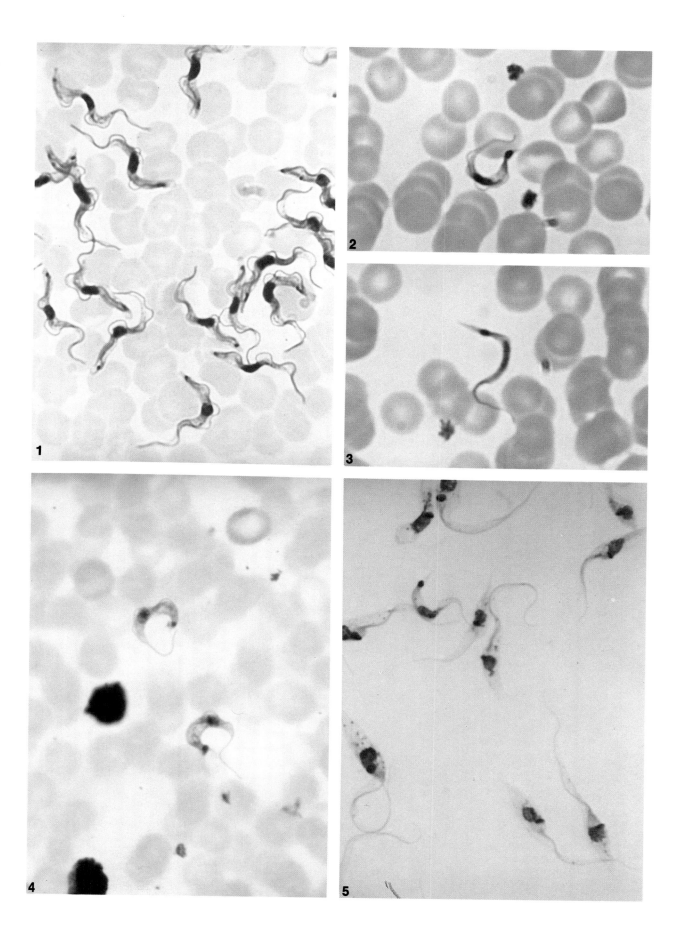

Leishmania species

CLASSIFICATION. — Flagellates.

DISEASES. — Cutaneous and visceral leishmaniases.

GEOGRAPHIC DISTRIBUTION. — Visceral leishmaniasis (kala-azar), caused by *Leishmania donovani*, occurs in China, India, the Mediterranean coast, Middle East, Africa, and Latin America. *Leishmania tropica* and other species cause different forms of cutaneous leishmaniases in tropical Africa, the Mediterranean area, Middle East, and in the western hemisphere.

LOCATION IN HOST. — *L. donovani* invades visceral organs after a primary lesion develops in the skin. The liver, spleen, and other components of the reticuloendothelial system primarily are involved. Cutaneous leishmaniasis is confined to lesions of the skin and lymphatics, although *L. braziliensis* of Central and South America causes mucocutaneous disease (espundia).

MORPHOLOGY. — Amastigotes. — These organisms are small and ovoid, measuring $1-5\,\mu$ long by $1-2\,\mu$ wide. They have a large nucleus, a rod-shaped parabasal body, and a small kinetoplast that gives rise to a short, internal flagellum.

Promastigotes. — This is the multiplication stage that normally occurs in the sandfly intermediate host. It is elongate, with a large nucleus and a kinetoplast at the anterior end that gives rise to a short, free, anteriorly directed flagellum. Inoculating culture media with amastigote stages from skin will produce these promastigote forms.

LIFE CYCLES. — Sandflies of the genera *Phlebotomus* and *Lutzomyia* are the arthropod intermediate host. The amastigote stages are ingested when the fly feeds on infected skin. In the midgut of the fly, the amastigotes transform into promastigotes, which then undergo multiplication. The promastigote is the infective stage inoculated when the fly feeds again.

DIAGNOSIS. — In endemic areas of cutaneous leishmaniasis, diagnosis usually is made on clinical grounds. Amastigote stages may be identified in aspirated material from the edges of cutaneous ulcers, or this material may be inoculated into culture media to produce promastigote stages. Clinical diagnosis of visceral leishmaniasis is sometimes difficult, and demonstration of organisms in spleen or liver may be required. Aspirates of material from bone marrow, blood, liver, or spleen can be inoculated into culture media and may reveal the infection by demonstration of promastigotes. Serologic techniques also may be helpful.

Diagnostic problems. — Amastigote stages may be difficult to detect in impression smears or in biopsy material. Amastigotes of *Leishmania* must be differentiated from organisms such as *Toxoplasma* and *Histoplasma*.

FIGURES. — Plate 24: 2-4.

PLATE 24

Trypanosoma cruzi and *Leishmania* species, Amastigotes and Promastigotes

Fig 1. *T. cruzi*, hematoxylin-eosin-stained rat skeletal muscle section. These are the intracellular amastigote forms, those in which this species multiplies in the mammalian host. Note the bar-shaped kinetoplast that can be seen close to the spherical nucleus in many of the parasites. This structure clearly differentiates this protozoan from organisms such as *Toxoplasma*, *Histoplasma*, and *Sarcocystis*.

Fig 2. *L. braziliensis*, Giemsa-stained human skin ulcer smear. The amastigotes shown here range in shape from spherical to elongate. The smaller, dark-staining kinetoplast can be seen alongside the spherical nucleus in most of the parasites. This is the only stage of the organism found in the mammalian host.

Fig 3. *L. donovani*, Giemsa-stained amastigotes in human bone marrow smear. These amastigotes (Leishman-Donovan bodies) are morphologically indistinguishable from those of *L. braziliensis*, seen in Figure 2.

Fig 4. *L. tropica*, Giemsa-stained promastigotes in culture. Although not all parasites are in the same focal plane, two at the 9-o'clock position are in sharp focus. Note that the flagellum arises near the kinetoplast and extends from the anterior end of the organism. There is no undulating membrane. Diagnosis of leishmaniasis frequently is made by inoculation of culture media with aspirates from skin, bone marrow or biopsy materials, and subsequent recovery of these promastigote stages.

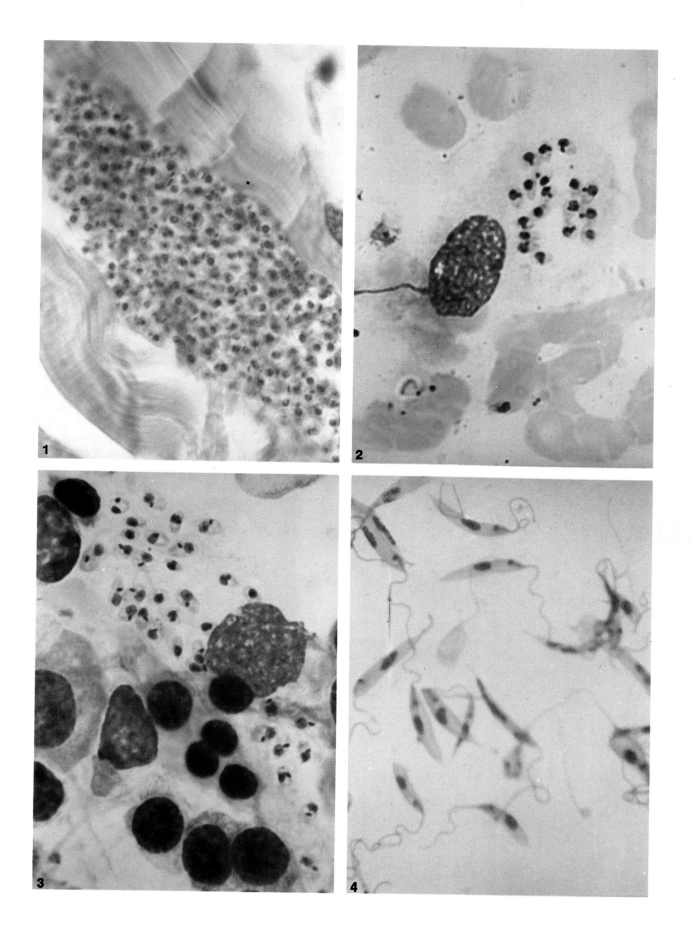

Pneumocystis carinii

CLASSIFICATION. — Protozoa (?)

DISEASE. — Pneumocystosis; interstitial plasma cell pneumonia.

GEOGRAPHIC DISTRIBUTION. — Cosmopolitan.

LOCATION IN HOST. — Lung tissue.

MORPHOLOGY. — Trophozoites. — Small, $1-5$ μ in diameter, with ovoid or ameboid appearance.

Cysts. — The cysts are $3.5-7.0$ μ in diameter, have an outer membrane of variable thickness, and usually contain four to eight nuclei, each of which measures $1-2$ μ in diameter.

LIFE CYCLE. — Although the life cycle has not been clearly elucidated, Pneumocystis probably exists in the environment and in the lungs of rodents and humans without causing apparent disease. However, in hosts that are immunosuppressed for extended periods of time, or in premature and malnourished infants, the organism may proliferate in lung tissue, leading to consolidation, dyspnea, and even death.

DIAGNOSIS. — Usually established by demonstrating organisms in lung tissue. Specimens obtained by lung biopsy, lung aspirates, or bronchial brushings are more satisfactory for examination than are sputum or bronchial washings because more organisms are likely to be detected. Although a variety of stains (Giemsa's, methylene blue, hematoxylin-eosin) have been used, the Grocott modification of the Gomori methenamine-silver nitrate stain is especially useful because organisms stain darkly and are readily identifiable against an unstained background.

Diagnostic problems. — Obtaining a suitable specimen for examination is essential. Pneumocystis must be differentiated from yeasts, fungi, and other organisms.

FIGURES. — Plate 25: 1 – 4.

PLATE 25

Pneumocystis carinii, Cysts

Fig 1. Cysts in human lung section, Gomori methenamine-silver nitrate stain. The cysts contrast sharply with the pale green background. Many of the cysts have a cup-shape appearance due to partial collapse of the cyst wall. Dark spots may represent wrinkles in the cyst wall or the "opercular-like" structure described by investigators.

Fig 2. The cysts in this human lung impression smear stained with Gomori methenamine-silver nitrate, are clearly evident. Note that many have a cup-shape.

Fig 3. Characteristic cysts stained with Gomori methenamine-silver nitrate are seen at high magnification.

Fig 4. In this hematoxylin-eosin-stained section of human lung, the alveoli appear to be filled with a pink-staining, foamy material. Neither cysts nor organisms can be seen when hematoxylin-eosin stain is used.

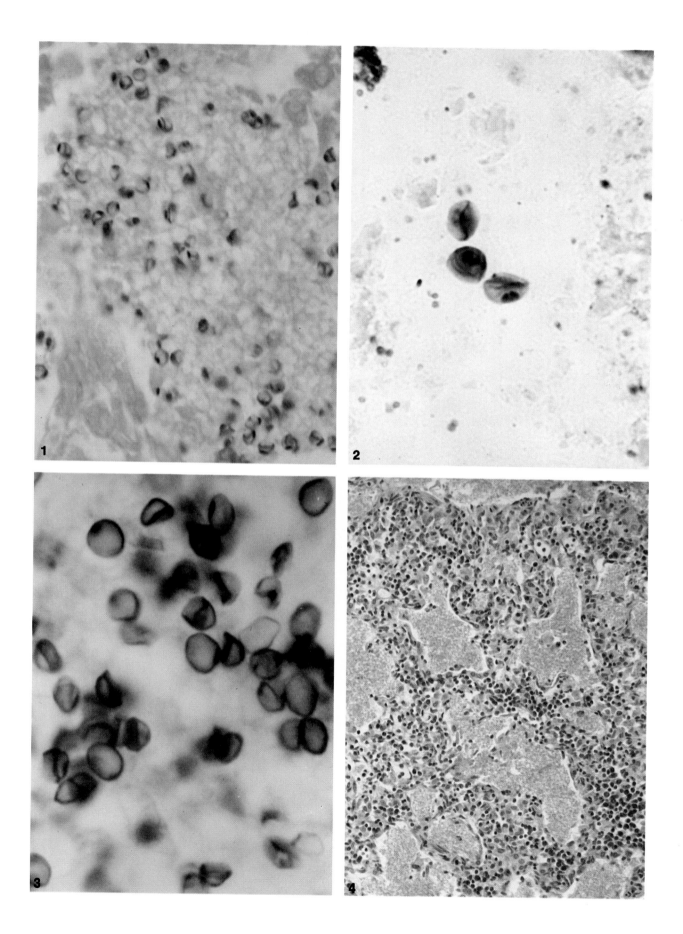

Enterobius vermicularis

CLASSIFICATION. — Nematode.

DISEASE. — Enterobiasis, pinworm infection, oxyuriasis.

GEOGRAPHIC DISTRIBUTION. — Cosmopolitan.

LOCATION IN HOST. — Cecum, appendix, colon, and rectum.

MORPHOLOGY. — Adult worms. — Males are 2.5 mm long by 0.1–0.2 mm wide, with blunt posterior ends and a single spicule. Females are 8–13 mm long by 0.3–0.5 mm wide, and have long pointed tails. In both sexes, there are cephalic inflations, and the esophagus is divided into three portions — a corpus, isthmus, and bulb.

Eggs. — Elongate, flattened on one side, with a thick, colorless shell, 50–60 μ by 20–30 μ. Eggs are partially embryonated when laid.

LIFE CYCLE. — Females usually emerge from anus at night and lay their partially embryonated eggs on the perianal surface. Eggs embryonate to the infective first stage within four to six hours. Infection usually is by direct transmission of eggs to mouth by hands or through fomites. Parasites develop in lower intestinal tract, and the prepatent period is three to four weeks. Adults normally live for only a few months.

DIAGNOSIS. — Eggs usually are detected in cellulose tape preparations applied to the patient's perianal region in the early morning prior to the patient's bathing or using the toilet. Eggs are sometimes found in fecal preparations. However, routine diagnosis by fecal examination is unreliable since eggs are not introduced into the fecal stream. Instead these are laid on the surface of the fecal material as it passes through the rectum. Not infrequently, adult females are seen around the anus or on the surface of stool specimens.

Diagnostic problems. — Since eggs are not usually found in routine fecal examinations, cellulose tape preparations are the most reliable means for detecting this infection.

COMMENTS. — The condition caused by *E. vermicularis* is a familial and group infection, more prevalent in children. It is very common in day nurseries and institutional settings.

FIGURES. — Plate 26:1–4; Plate 27:1–4.

PLATE 26

Enterobius vermicularis, Eggs

Fig 1. Three typical embryonated (infective) eggs are seen in a cellulose tape preparation. The shell appears moderately thick and hyaline. Note characteristic flattening of one side of the egg and the presence of a larva within.

Fig 2. This embryonated egg is oriented so that the flattened surface of the shell is not apparent.

Fig 3. This egg, found in a fecal smear preparation, is only partially embryonated. At this stage, the larva frequently is referred to as the "tadpole" stage.

Fig 4. When found in feces, the eggs are typically unembryonated. Note the undifferentiated contents of each.

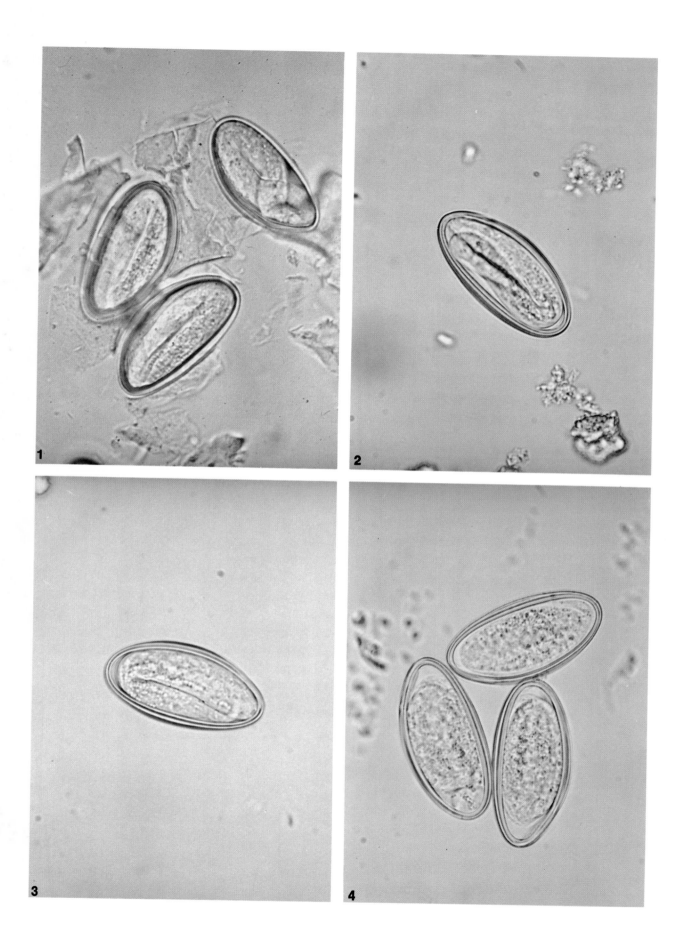

PLATE 27

Enterobius vermicularis, **Adult Worms**

Fig 1. Adult female worms are white, frequently seen in or on feces, and sometimes adhere to cellulose tape preparations. The female is small and has a long, sharply pointed tail. Note that the body of the worm is filled with eggs.

Fig 2. Contrary to the representation here, adult male worms are much smaller than females and are detected less frequently in feces or on cellulose tape preparations. Note that the male resembles the female except for its blunt, posterior end.

Fig 3. Adult female, anterior end. Adult pinworms can be identified by the characteristic dorsoventral cephalic inflations of the cuticle, and the shape and structure of the esophagus, ie, divided into muscular and bulbous portions separated by a short, narrow isthmus. Some of these same features also can be observed in Figures 1 and 2.

Fig 4. Hematoxylin-eosin-stained adult worm. Adult pinworms frequently are seen in histologic sections of the appendix. They may be identified in these cross sections by the presence of prominent lateral alae *(arrow)* on the surface of the cuticle.

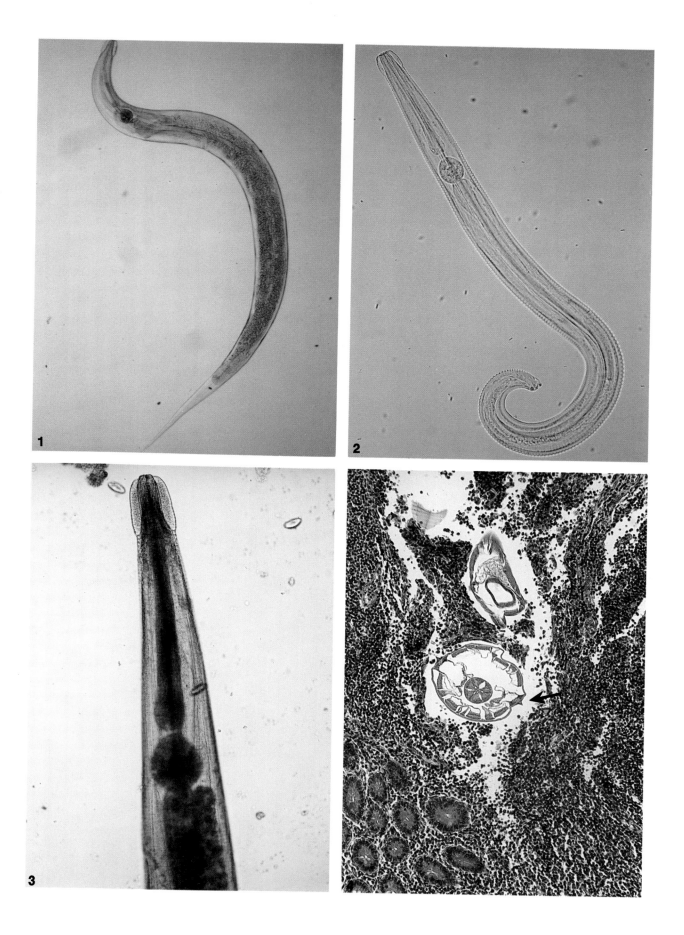

Ascaris lumbricoides

CLASSIFICATION. — Nematode.

DISEASE. — Ascariasis.

GEOGRAPHIC DISTRIBUTION. — Cosmopolitan, but more prevalent in warm, moist regions of the world.

LOCATION IN HOST. — Small intestine.

MORPHOLOGY. — Adult worms. — Males are 15–31 cm by 2–4 mm, and have a curved tail. Females are 20–35 cm by 3–6 mm, and have a straight tail.

Eggs. — Fertile eggs are bile-stained, have a mammillated, thick shell, measure 55–75 μ by 35–50 μ, and are in the one-celled stage when passed in feces. In some instances, the outer albuminoid, mammillated layer is absent (decorticated eggs). Infertile eggs are elongate, 85–95 μ by 43–47 μ, and have thin shells, with the mammillated layer varying from grossly irregular mammillations to a relatively smooth layer almost completely lacking mammillations. The internal contents are a mass of disorganized, highly refractive granules.

LIFE CYCLE. — Females are oviparous. Unembryonated eggs in feces pass into the soil where they undergo development for two to three weeks. Each will contain an infective, second-stage larva. When infective eggs are ingested, larvae emerge in the small intestine and undergo an obligatory migration for about eight to nine days through the liver and lungs. They undergo considerable growth to reach a length of 1 mm, and then return to the small intestine where they grow to maturity. The prepatent period is about two months. Adult worms live for a year or less.

DIAGNOSIS. — Demonstration of characteristic eggs in feces.

Diagnostic problems. — Ascaris eggs are produced in such large numbers by adult female worms that even when a single pair of worms is present, eggs can be detected by direct smear examination of feces. Although fertile eggs concentrate well by sedimentation or flotation concentration procedures, infertile eggs do not float with standard zinc sulfate solution (1.18 specific gravity). Consequently, these eggs may be missed if only a flotation concentration procedure is used for fecal examination. Infertile eggs sometimes may pose diagnostic problems, particularly if the outer mammillated layer is totally absent.

COMMENTS. — The disease associated with larval migration is referred to as Ascaris pneumonitis or Loeffler's syndrome. The erratic migration of adult worms to extraintestinal locations makes the presence of even one worm a potentially serious danger to the host.

FIGURES. — Plate 28:1–6; Plate 29:1–4.

PLATE 28

Ascaris lumbricoides, Unembryonated, Fertile Eggs

Fig 1. Note the characteristic bile-stained, mammillated external layer of the thick shell. The well defined ovum is in the one-cell stage, that normally found in fresh feces.

Figs 2–4. Although these three eggs are unembryonated and similar in most respects, they show considerable variation in the mammillations of the external layer.

Fig 5. In this otherwise normal egg, the external, mammillated layer is absent, and the egg is referred to as "decorticated." The shape of the egg and the thickness of the shell distinguish it from other smooth-shelled eggs, such as those of pinworm and hookworm, that also may be found in fecal specimens.

Fig 6. Both normal and decorticated eggs are seen in this illustration. Note the size difference between the egg with a mammillated layer and the decorticated egg. Other features are the same.

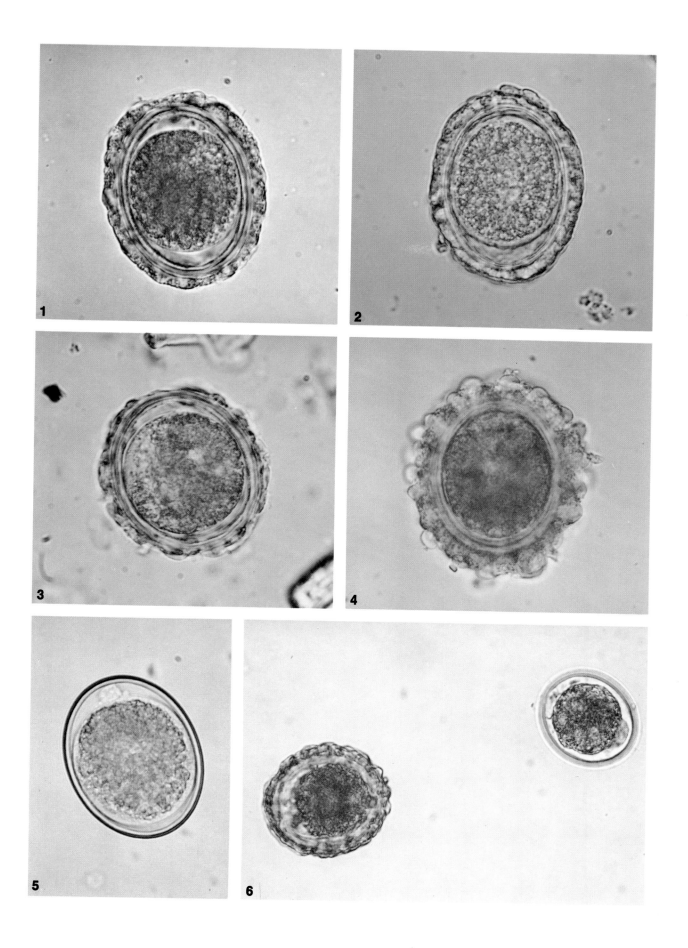

PLATE 29

Ascaris lumbricoides, Eggs

Fig 1.   This typical embryonated egg contains an infective, second-stage larva. These eggs normally develop in the soil. They are highly resistant to preservatives such as 10% formalin and may become infective when preserved in the laboratory. Care should be used in handling preserved eggs and adult worms. For technical reasons, the size of this egg is shown to be almost equal to that of the infertile eggs in Figures 2–4; however, note that infertile eggs are always larger than fertile eggs.

Fig 2.   Infertile eggs are larger, can have bizarre shapes, and may lack one or more of the normal shell layers. This egg has no mammillated layer. Also, note that the internal contents appear disorganized and globular.

Fig 3.   These two eggs pictured in the same field illustrate the range of variation that may be seen in infertile eggs.

Fig 4.   This infertile egg has a poorly developed, irregular mammillated layer.

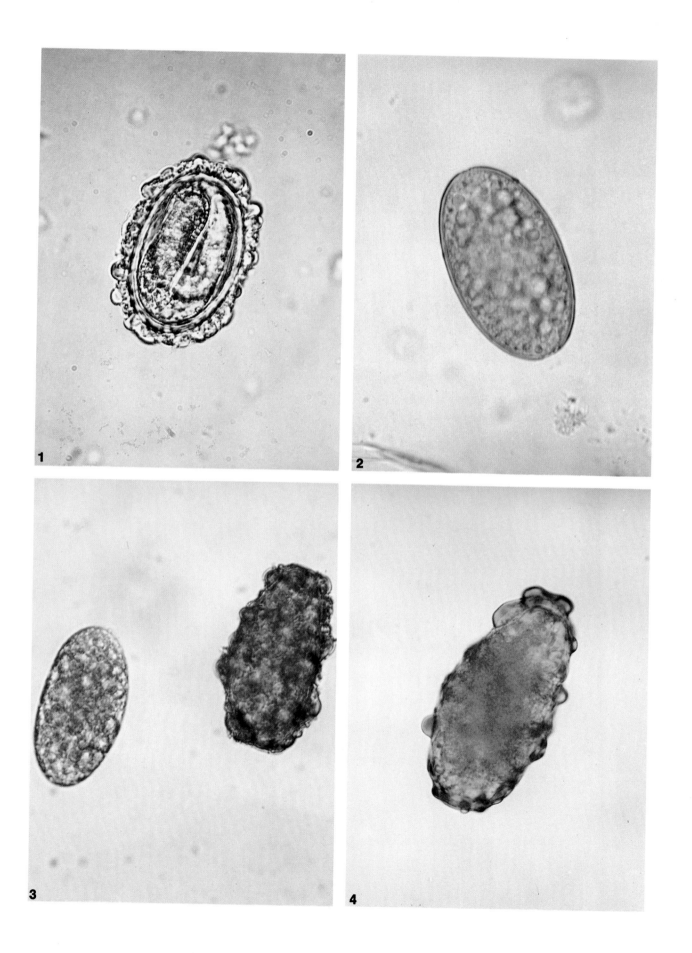

Trichuris trichiura

CLASSIFICATION. — Nematode.

DISEASE. — Trichuriasis, whipworm infection.

GEOGRAPHIC DISTRIBUTION. — Cosmopolitan, especially prevalent in warm, moist regions of the world.

LOCATION IN HOST. — Large intestine, cecum, and appendix.

MORPHOLOGY. — Adult worms. — Males are 30–45 mm long, with coiled posterior ends. Females are 35–50 mm, with straight posterior ends. Adults have long, slender, whip-like anterior ends, and thicker, short posterior ends. The esophagus is a slender tube surrounded by cells (stichocytes) collectively termed a stichosome.

Eggs. — Eggs are 50–55 μ by 22–24 μ and barrel-shaped, with bile-stained, thick shells and clear, mucoid "plugs" at each end. Eggs are unembryonated when passed.

LIFE CYCLE. — Females are oviparous. Eggs pass into soil where they undergo development for two to three weeks, and then will contain an infective, first-stage larva. When infective eggs are ingested, larvae emerge in the intestine and migrate to the large intestine where they develop to maturity. The prepatent period is approximately three months. The anterior end of adult worms is threaded into the mucosal epithelium. Adult worms may live for up to ten years or longer.

DIAGNOSIS. — Demonstration of characteristic eggs in feces.

Diagnostic problems. — Trichuris eggs are readily recognizable and usually pose no problems. In light infections, eggs may be difficult to find in fecal examinations unless concentration procedures are used. In patients treated with some anthelmintics, aberrant eggs may be passed in feces, or occasionally very large eggs (80 μ or more) can be produced. In human infections with the dog whipworm, T. vulpis, the eggs always are much larger (72–90 μ by 32–40 μ) and more barrel-shaped than those of T. trichiura.

FIGURES. — Plate 30:1–6; Plate 31:1–6.

PLATE 30

Trichuris trichiura, Adults and Eggs

Fig 1. This illustration demonstrates the gross features of male and female adult worms. Note that both are whip-like, and that the male, in contrast to the female, has a characteristically coiled tail. These worms, though small, are readily seen but rarely found in feces, except following treatment.

Figs 2–4. Note the gross features of the typical, unembryonated, fertile egg, ie, the bile-stained, thick, smooth shell with polar prominences, often called polar "plugs."

Fig 5. This embryonated egg has undergone development for two to three weeks and now contains an infective larva. These eggs are never seen in stool specimens and normally are found only in soil.

Fig 6. Helminth eggs, including Trichuris, often are seen in stained fecal smears. Note the typical features of this unembryonated, fertile Trichuris egg in a trichrome-stained preparation.

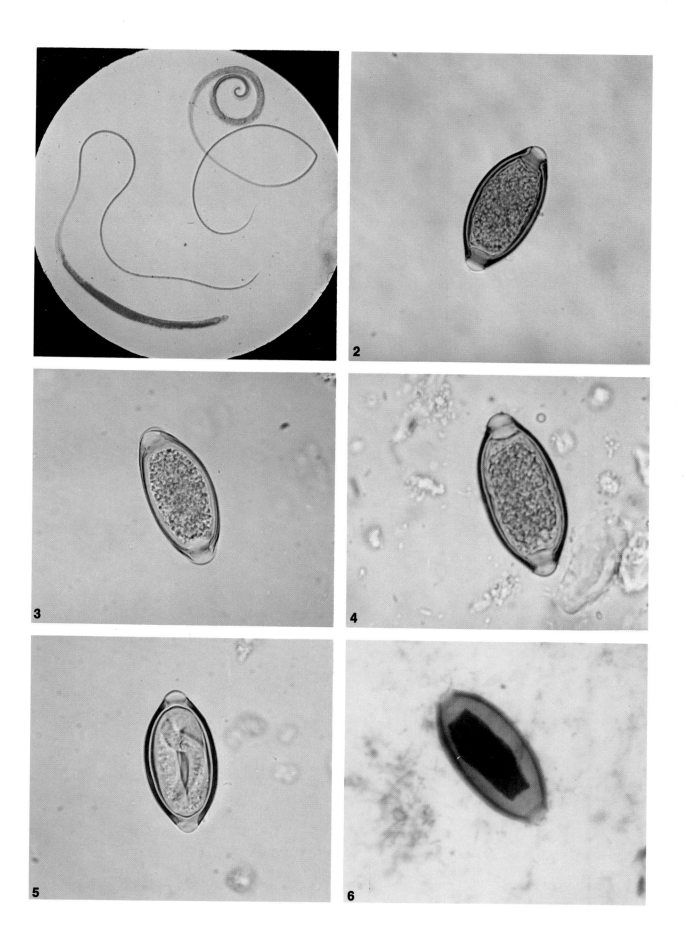

2

3

4

5

6

PLATE 31

Trichuris trichiura and *T. vulpis,* Eggs

Fig 1. *T. trichiura,* unembryonated, fertile egg. A typical egg seen in fresh feces is shown for comparison in size and structure with the other eggs in this plate.

Fig 2. *T. vulpis,* unembryonated, fertile egg. This is the egg of the dog whipworm. Note that the egg of this species, occasionally found in humans, is much larger and broader than *T. trichiura.*

Fig 3. This *T. trichiura* egg, often seen in the early stages of patent infections, appears abnormal, but has many features of a normal egg that help identify it.

Figs 4–6. *T. trichiura,* abnormal eggs. Following treatment with some anthelmintics, abnormal eggs often are found in feces.

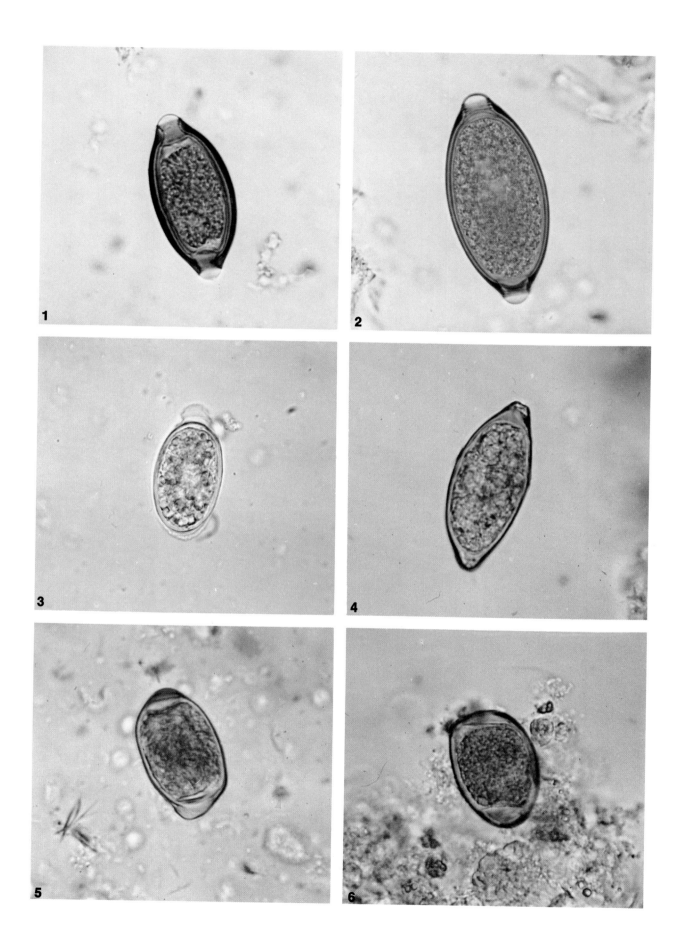

Capillaria philippinensis and C. hepatica

CLASSIFICATION. — Nematode.

DISEASES. — Intestinal capillariasis (*C. philippinensis*); hepatic capillariasis (*C. hepatica*).

GEOGRAPHIC DISTRIBUTION. — Philippines and Thailand (*C. philippinensis*); cosmopolitan (*C. hepatica*).

LOCATION IN HOST. — Intestine (*C. philippinensis*); liver (*C. hepatica*).

MORPHOLOGY. — Adult worms. — *C. philippinensis* adult males are 2.3–3.2 mm long by 30 μ wide. Females are 2.5–4.3 mm by 47 μ. Males have a single spicule. Females may contain in utero unembryonated or embryonated eggs and occasionally larvae. *C. hepatica* adults may reach lengths of up to 20 mm, but are rarely — if ever — seen intact, since they mature and ultimately die in the parenchyma of the liver.

Eggs. — The eggs of *C. philippinensis* are 36–45 μ long by 21 μ wide, have a striated shell, and inconspicuous polar prominences at each end. They usually are unembryonated when passed in feces. The eggs of *C. hepatica*, which are 51–67 μ long by 30–35 μ wide, have a distinctly striated shell, the polar prominences are shallow, and they are unembryonated when found in feces.

LIFE CYCLES. — *C. philippinensis* eggs embryonate in water and are ingested by small fish, within which the larvae hatch and migrate to the mesenteries to become infective in one to two weeks. Fish-eating birds appear to be the normal definitive host for this parasite. Human infections occur when fish are eaten raw. Internal autoinfection is a normal feature of the life cycle in mammalian hosts, and may result in overwhelming numbers of parasites that cause death of the host.

Rodents are the usual host for *C. hepatica* infection. Adult worms live for only one to four months after the worms mature in the liver parenchyma, and the female lays her eggs directly into the tissue. These eggs remain unembryonated until the liver is eaten by another animal, and the eggs are digested free and then excreted in feces. (Thus, the passage of *C. hepatica* eggs in feces is indicative of spurious infection.) Eggs pass into the soil, embryonate to the infective stage, and rodents, other mammals, and humans acquire infection by eating these eggs. Larvae hatched from the eggs migrate to the liver and reach maturity within two to three weeks. Egg-laying commences by the third week.

DIAGNOSIS. — *C. philippinensis* is diagnosed by demonstration of characteristic eggs in feces. *C. hepatica* is a form of visceral larva migrans, and, since eggs are not passed in the feces of those who actually have the infection, it is difficult to diagnose the infection. Liver biopsy perhaps is the most reliable method available.

Diagnostic problems. — As just mentioned, *C. hepatica* is difficult to diagnose because of the nature of the life cycle.

FIGURES. — Plate 32:1–6.

PLATE 32

Capillaria philippinensis and C. hepatica, Eggs

Figs 1 and 2. *C. philippinensis*, unembryonated fertile eggs. These small eggs have striated shells and inconspicuous polar prominences.

Figs 3 and 4. *C. hepatica*, unembryonated, fertile eggs. These eggs, though similar in gross appearance to *Trichuris trichiura* eggs, have thick, striated shells, and polar prominences. When seen in human feces, they are indicative of spurious infection.

Figs 5 and 6. *C. hepatica*, eggs, in liver, hematoxylin-eosin stain. In their natural rodent hosts, as well as in humans, adult worms live in the liver parenchyma and lay their eggs directly in the tissue. Their morphologic features are evident in these two tissue sections.

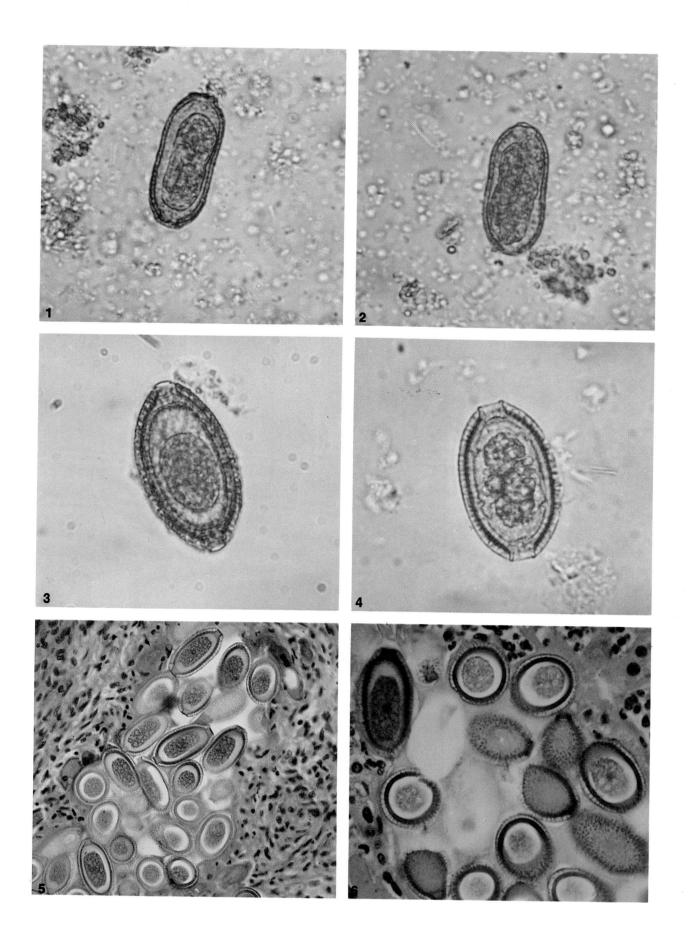

Trichostrongylus species

CLASSIFICATION. — Nematode.

DISEASE. — Trichostrongylosis.

GEOGRAPHIC DISTRIBUTION. — Worldwide, especially in rural areas where herbivorous animals are raised.

LOCATION IN HOST. — Small intestine.

MORPHOLOGY. — Adult worms. — Small, slender worms, with males 2 – 8 mm long by 50 – 60 μ in diameter and females 3 – 9 mm long by 55 – 80 μ. Males have a bursa.

Eggs. — Thin-shelled, colorless, measuring 75 – 95 μ by 40 – 50 μ. Eggs taper at one end, and the inner membrane frequently is wrinkled. The ovum is in advanced cleavage when passed in feces.

LIFE CYCLE. — Eggs are shed in feces into soil, where they undergo development and hatch as first-stage larvae within several days. Larvae undergo development to the infective third stage, and infection is acquired by ingestion of these larvae. The prepatent period is approximately one month.

DIAGNOSIS. — Demonstration of characteristic eggs in feces.

Diagnostic problems. — These eggs sometimes are confused with hookworm eggs, but trichostrongyle eggs are considerably larger, and have one tapered end.

COMMENTS. — Although most species of *Trichostrongylus* are parasites of herbivorous animals, humans may acquire infection with various species — *T. orientalis*, *T. colubriformis*, *T. axei*, and others. This infection is especially common in persons from rural areas of Middle Eastern countries, such as Iran and Iraq.

FIGURES. — Plate 33: 1 – 3; also see Plate 34: 6.

PLATE 33

Trichostrongyle and Hookworm Eggs

Figs 1–3. The eggs of trichostrongyles (*Trichostrongylus* spp) are large, have a thin, hyaline shell, and the germinal mass is in an advanced stage of development in fresh feces. Note that the egg is elongate and tapered at one end and the germinal mass does not fill the shell.

Fig 4. For purposes of comparison with the trichostrongyle eggs, note that the typical hookworm egg is smaller, bluntly rounded at the ends, and, characteristically, is in the 4 – 8 cell stage of development in fresh feces.

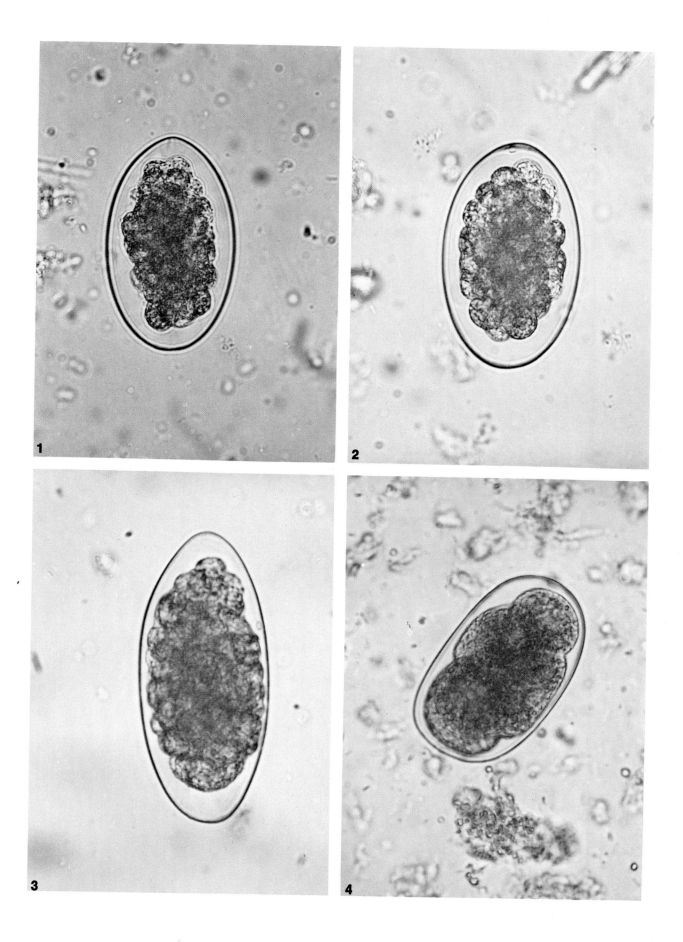

Necator americanus

CLASSIFICATION. — Nematode.

DISEASE. — Hookworm infection; ankylostomiasis.

GEOGRAPHIC DISTRIBUTION. — Western hemisphere, Central and South Africa, southern Asia, South Pacific, and India.

LOCATION IN HOST. — Small intestine.

MORPHOLOGY. — Adult worms. — Males measure 7–9 mm by 0.3 mm and are bursate with two spicules that fuse at their distal end. Females are 9–11 mm by 0.4 mm. Adults have a buccal capsule containing cutting plates rather than teeth.

Eggs. — Thin-shelled, colorless, and measure 60–75 μ by 36–40 μ; usually in early cleavage when passed in feces.

Larvae. — First-stage rhabditoid larvae that hatch from eggs are 250–300 μ long by 17 μ. They have a long buccal canal and the genital primordium is small and difficult to see. Infective, third-stage, filariform larvae are 580 to 620 μ long. These have a pointed tail and a ratio of esophageal-to-intestinal length of 1:4. The sheath about the larva is conspicuously striated.

LIFE CYCLE. — Females are oviparous. Eggs are shed in feces into soil where they embryonate and hatch in approximately 24 hours. Larvae reach the infective third stage in approximately one week. Human infection is acquired by skin penetration by these larvae. Larvae undergo some growth in lung prior to migration to the small intestine. The prepatent period is about five to six weeks. Adults may live up to 15 years, but the usual life span is between five to ten years.

DIAGNOSIS. — Demonstration of eggs in feces.

Diagnostic problems. — Eggs of this species are indistinguishable from those of *Ancylostoma duodenale*. If eggs hatch in feces due to delay in fecal examination, these first-stage larvae must be differentiated from those of *Strongyloides stercoralis*, which normally are passed in feces. Whereas hookworm first-stage larvae have a long buccal canal and an inconspicuous genital primordium, the larvae of *Strongyloides* have a short buccal canal and a prominent genital primordium.

FIGURES. — Plate 34: 1–5; Plate 35: 1 and 3; Plate 36: 1 and 2; also see Plate 33: 4.

PLATE 34

Hookworm and Trichostrongyle Eggs

Figs 1, 2, and 4. Hookworm, fertile eggs. Hookworm eggs typically have a thin, hyaline shell, and are bluntly rounded at the ends. They usually are in the 4–8 cell stage in fresh feces, and the developing ovum tends to fill the shell. In feces, it is not possible to differentiate the eggs of the human hookworms *Necator* and *Ancylostoma*. The egg in Figure 2 is reproduced at a slightly higher magnification than those in Figures 1 and 4.

Fig 3. Hookworm, embryonated egg. As illustrated here, eggs in feces left standing at room temperature for at least a day may develop, become fully embryonated, and contain first-stage larvae.

Fig 5. Hookworm and *Trichuris trichiura* eggs. For purposes of comparison, note, in addition to the typical morphologic features, the size difference between hookworm and *Trichuris* eggs. People frequently are infected with both of these parasites simultaneously.

Fig 6. This trichostrongyle egg illustrates the size and morphologic differences between hookworm and trichostrongyle eggs.

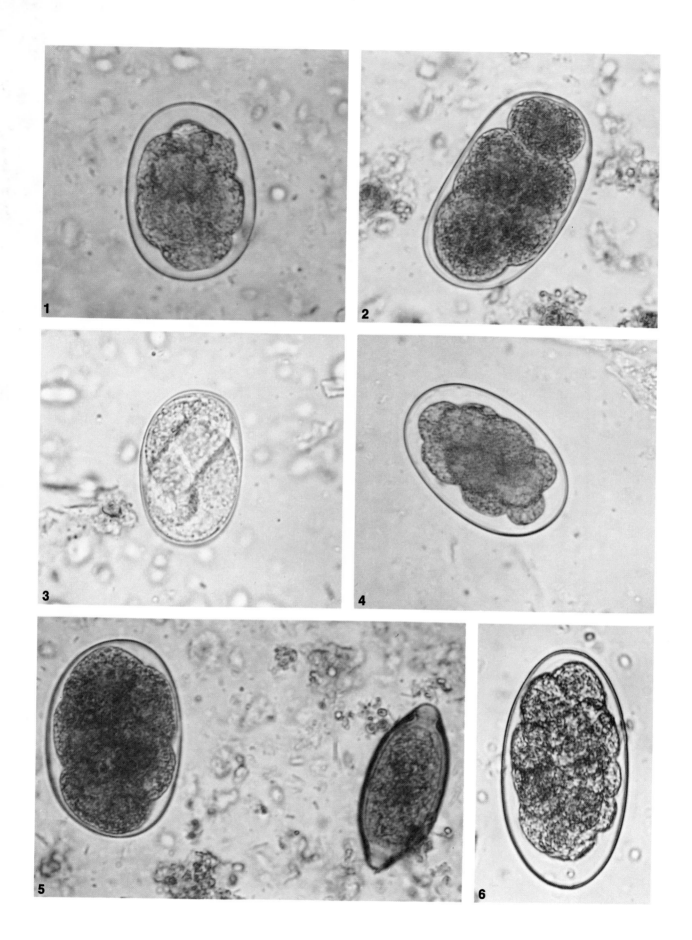

Ancylostoma duodenale

CLASSIFICATION. — Nematode.

DISEASE. — Hookworm infection; ankylostomiasis.

GEOGRAPHIC DISTRIBUTION. — Southern Europe, northern parts of Africa, China, India, and Japan, and, sporadically, in other areas where infection has been introduced more recently.

LOCATION IN HOST. — Small intestine.

MORPHOLOGY. — Adult worms. — Males measure 8–11 mm by 0.4–0.5 mm, and are bursate, with two spicules that do not fuse at their distal ends. Females measure 10–13 mm by 0.5–0.7 mm. Adults have a buccal capsule containing two pairs of teeth.

Eggs. — Thin-shelled and colorless, measuring 55–65 μ by 36–40 μ.

Larvae. — First-stage rhabditoid larvae that hatch from eggs are 250–350 μ long by 17 μ. They have a long buccal canal and the genital primordium is small and difficult to see. Infective, third-stage, filariform larvae are 625–675 μ long. They have a pointed tail and a ratio of esophageal length to intestinal length of 1:4. The sheath is not as conspicuously striated as in *Necator*.

LIFE CYCLE. — Females are oviparous. Eggs are shed in feces on soil where they embryonate and hatch in approximately 24 hours. Larvae reach the infective third stage in approximately one week, and human infection is obtained both by mouth and by direct skin penetration. Larvae undergo no growth if they pass through lung prior to reaching maturity in the small intestine. The prepatent period is five to six weeks. Adults usually live for at least five to ten years.

DIAGNOSIS. — Demonstration of eggs in feces.

Diagnostic problems. — As described for *Necator americanus*.

COMMENTS. — Since hookworm species cannot be differentiated on the basis of their eggs, it is necessary to culture larvae or to recover adult worms for specific diagnosis. Stool specimens must not be refrigerated before attempting to culture larval stages, as *Necator* is especially sensitive to cold.

FIGURES. — Plate 35:1 and 3; Plate 36:1 and 2; also see Plate 33:4 and Plate 34:1–5.

PLATE 35

Iodine-stained Hookworm and *Strongyloides stercoralis* Larvae in Feces

Figs 1 and 3. Hookworm first-stage (rhabditoid) larvae. These larvae may be confused with *Strongyloides* larvae, normally passed in feces. The hookworm larva has a long, tubular, buccal canal and the genital primordium cannot be seen.

Figs 2 and 4. *S. stercoralis*, first-stage (rhabditoid) larvae. The rhabditoid larva of *Strongyloides* is approximately the same size as that of the hookworm first-stage larva. However, the former has a short, tubular, buccal canal, and the genital primordium is prominent and easily seen lying between the midpoint of the intestine and the ventral body wall *(arrow)*. Note that *Strongyloides* first-stage larvae are the diagnostic stage for this infection, whereas hookworm larvae will be seen in feces only when left at room temperature for one or more days.

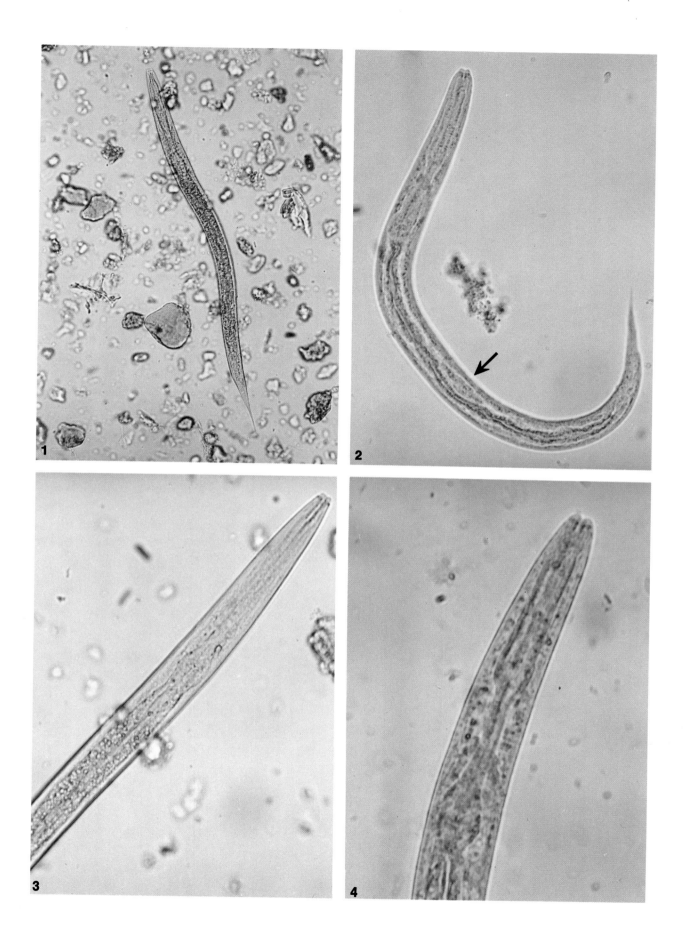

Strongyloides stercoralis

CLASSIFICATION. — Nematode.

DISEASE. — Strongyloidiasis.

GEOGRAPHIC DISTRIBUTION. — Cosmopolitan, but more prevalent in warm climates and areas where the water table is high.

LOCATION IN HOST. — Small intestine.

MORPHOLOGY. — Adult worms. — There is no parasitic male. Parasitic females are thread-like, 2.1–2.7 mm long by 30–40 μ.

Eggs. — They are thin-shelled and resemble hookworm eggs in appearance and size. These eggs rarely are seen, inasmuch as they embryonate and hatch in the mucosa of the small intestine.

Larvae. — First-stage rhabditoid larvae are 180–380 μ long by 14–20 μ wide. They have a short buccal canal and a prominent genital primordium. Infective, third-stage, filariform larvae are approximately 500 μ long. The tail is notched and the ratio of esophageal length to intestinal length is 1:1. There is no sheath about infective larvae, as in the hookworms.

LIFE CYCLE. — Females are ovoviviparous. Rhabditoid larvae hatch in the mucosal epithelium of the small intestine, enter the lumen, and then pass out in feces to undergo further development in the soil. Infective, third-stage, filariform larvae may develop in soil directly, or there may be production of a free-living generation that has both male and female worms. Free-living females will produce eggs that embryonate and hatch and become third-stage larvae.

Human infection is acquired through skin penetration by third-stage filariform larvae. Parasitic females are parthenogenetic, and mature in the mucosal epithelium of the small intestine in approximately one month, at which time they begin to produce embryonated eggs. It is possible for first-stage larvae to develop to third-stage larvae within the intestinal tract of debilitated or immunosuppressed persons, and initiate internal autoinfection. In these patients, *Strongyloides* may be a fulminating and fatal infection.

DIAGNOSIS. — Demonstration of characteristic first-stage larvae in feces.

Diagnostic problems. — In areas where this parasite and hookworm infections both exist, it sometimes may be necessary to differentiate *Strongyloides* larvae from hookworm larvae, which can hatch in stool specimens if examination is delayed. *Strongyloides* larvae have a short buccal canal and prominent genital primordium, whereas, in hookworms, the buccal canal is long and the genital primordium is inconspicuous. The number of *Strongyloides* larvae frequently may be very scant in feces. Consequently, many infections go undiagnosed unless special concentration procedures are used.

COMMENTS. — It cannot be emphasized too greatly that *Strongyloides* infections frequently are latent. The cessation or infrequent passage of larvae in chronic, often asymptomatic, infections may make diagnosis of this parasite difficult unless many large samples of feces are examined. Patients selected for immunosuppressive therapy, especially those from areas where *Strongyloides* infection is endemic, should be carefully screened for this infection prior to initiation of therapy.

FIGURES. — Plate 36:3 and 4; Plate 37:1–4; also see Plate 35:2 and 4.

PLATE 36

Hookworm and *Strongyloides* Larvae, in Fecal Culture

Fig 1. Hookworm, third-stage, filariform (infective) larva. The third-stage larva is much longer and more slender than its rhabditoid stage, has a short esophagus, a long intestine, and a sharply pointed tail *(inset)*. The iodine stain used here demonstrates the relative lengths of the esophagus and intestine — a ratio of approximately 1:4 in hookworm larvae. This stage normally is found in soil or in feces cultured for five days or longer. It is possible to differentiate species of the genera *Necator* and *Ancylostoma* based on the morphology of third-stage larvae.

Fig 2. The morphologic and diagnostic features of the infective, third-stage, hookworm larva are evident — even when left unstained.

Figs 3 and 4. *Strongyloides* third-stage, filariform (infective) larvae. The filariform larvae of *S. stercoralis*, as in hookworms, are longer and more slender than their rhabditoid stage. With iodine stain (Fig 3) or unstained (Fig 4), the larva is easily differentiated from that of hookworm by its long esophagus, equal in length to the intestine, and by its blunt, notched tail (Fig 4, *inset*). This filariform larva normally occurs only in soil or in cultured feces.

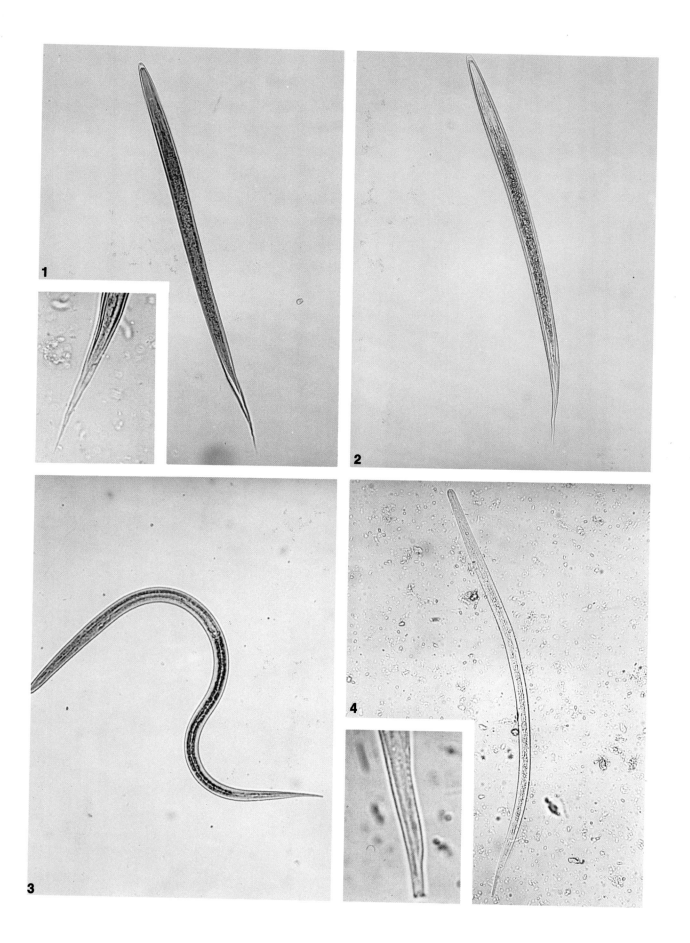

PLATE 37

Strongyloides stercoralis, Adult Worms

Fig 1.   Rhabditoid larvae *(arrow)* may develop in soil or in cultured feces into free-living adult worms. The unstained, free-living, adult male worm illustrated here can be recognized by its reproductive structures and its characteristically curved, short, pointed tail.

Fig 2.   This unstained, free-living adult female is larger than the male and usually is recognized by the presence of developing eggs in the reproductive tract. In contrast to that of the male, the pointed tail is not curved.

Fig 3.   Free-living adult female, iodine stain. If feces are allowed to stand at room temperature and remain moist for two to three days, the rhabditoid larvae *(arrow)* may develop into free-living adult worms.

Fig 4.   The parasitic, adult female of *Strongyloides* lives threaded into the mucosal epithelium of the human small intestine. There is no parasitic male; the female reproduces by parthenogenesis. In contrast to the free-living adults, the female appears small, very slender, and filariform, and has a short, pointed tail. Parasitic females rarely are seen in feces.

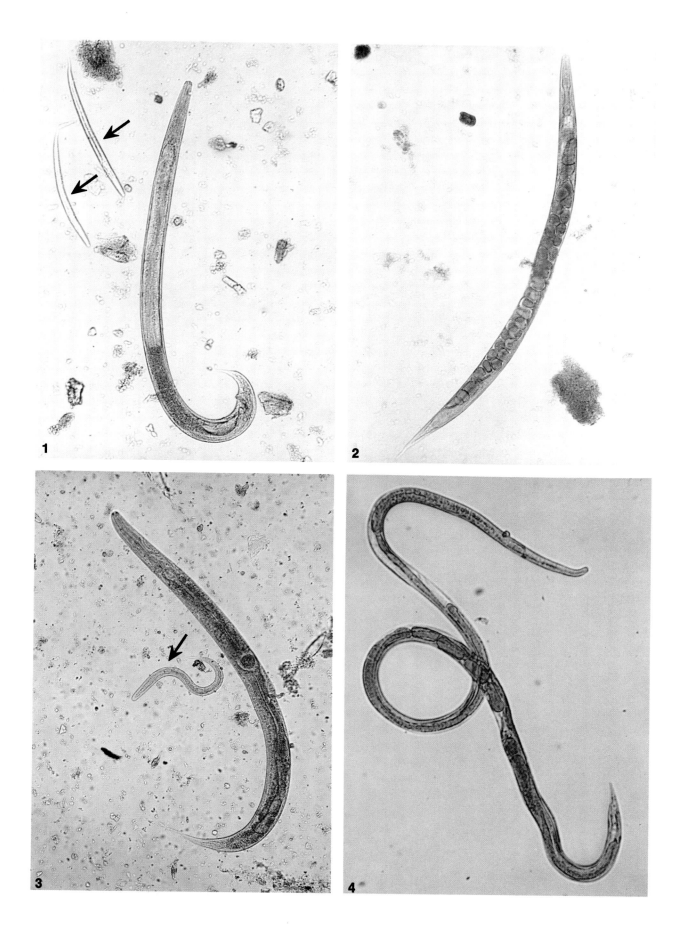

Trichinella spiralis

CLASSIFICATION. — Nematode.

DISEASE. — Trichinosis.

GEOGRAPHIC DISTRIBUTION. — Cosmopolitan, but more prevalent in Europe and North America than in tropical countries.

LOCATION IN HOST. — Adult worms live in intestinal tract for several weeks to several months. The larval stage, encapsulated in muscle tissue, may live for several years.

MORPHOLOGY. — Adult worms. — The adult male is minute, measuring 1.4 – 1.6 mm in length by 40 – 60 μ, and has two large, fleshy papillae at the posterior end. Adult females are 2.5 – 4.0 mm long by 100 – 150 μ wide. The reproductive tract is filled with developing eggs and larvae. In both sexes, the esophagus *(stichosome)* consists of a thin, narrow tube surrounded by a column of glandular cells called *stichocytes*.

Larvae. — Minute larvae, produced by the adult female, measure 100 μ long by 6 μ in diameter. Infective larvae in muscle tissue measure 0.8 – 1.0 mm long.

LIFE CYCLE. — When infective larvae in muscle tissue are ingested, they reach maturity in the intestine in approximately one week. Adult females deposit their minute larvae in the mucosal epithelium. These larvae enter the bloodstream and are carried to muscle tissue throughout the body. The larvae grow in muscle tissue, become infective in approximately one month, and are encapsulated by host tissue. All mammals may be infected by this parasite, but pigs and rats are the most important reservoir hosts in nature.

DIAGNOSIS. — Usually based on clinical symptoms and history of ingestion of poorly cooked meat, especially pork products. Serologic tests are useful. Although muscle biopsies usually are not performed to detect larvae, the procedure may be used. Diagnosis of animal infections is best established by examination of tissue. Larvae are most abundant in tongue, masseter muscle, diaphragm, and other active muscle tissues.

Diagnostic problems. — In light infections, symptoms may be vague and trichinosis not considered in a differential diagnosis. In animal infections, a few press preparations of muscle tissue may not be adequate to demonstrate larvae. Thus, artificial digestion of large amounts of muscle tissue may be required.

COMMENTS. — Most infections in the United States now are attributable to eating of poorly cooked or smoked pork products derived from hogs raised on private farms, rather than from pork processed by commercial meat packing firms. Bear meat has been shown to be infected, and numerous cases of human infection have been traced to this source.

FIGURES. — Plate 38:1 – 4.

PLATE 38

Trichinella spiralis, Adult Worms and Larvae

Fig 1.   Adult female. This stage inhabits the intestinal mucosa, is small and slender, with a tapered anterior end and a bluntly rounded posterior end. This parasite is most readily identified by the presence of the characteristic stichosome. The female reproductive tract is filled with developing eggs and larvae. Adult worms are short-lived, and rarely may be accidental findings in the feces of recently infected persons.

Fig 2.   The adult male also is found in the intestine and can be differentiated from the female by its smaller size and the presence of two conspicuous papillae at the posterior end.

Figs 3 and 4.   The larvae produced by adult females migrate via the bloodstream to the musculature where they undergo considerable growth and development, and become encapsulated by host tissue. Larvae may be found in press preparations of skeletal muscle (Fig 3), or in hematoxylin-eosin-stained histologic sections of the same tissues (Fig 4). In this stained section of diaphragm (Fig 4), note the presence of four larvae surrounded by the host tissue reaction.

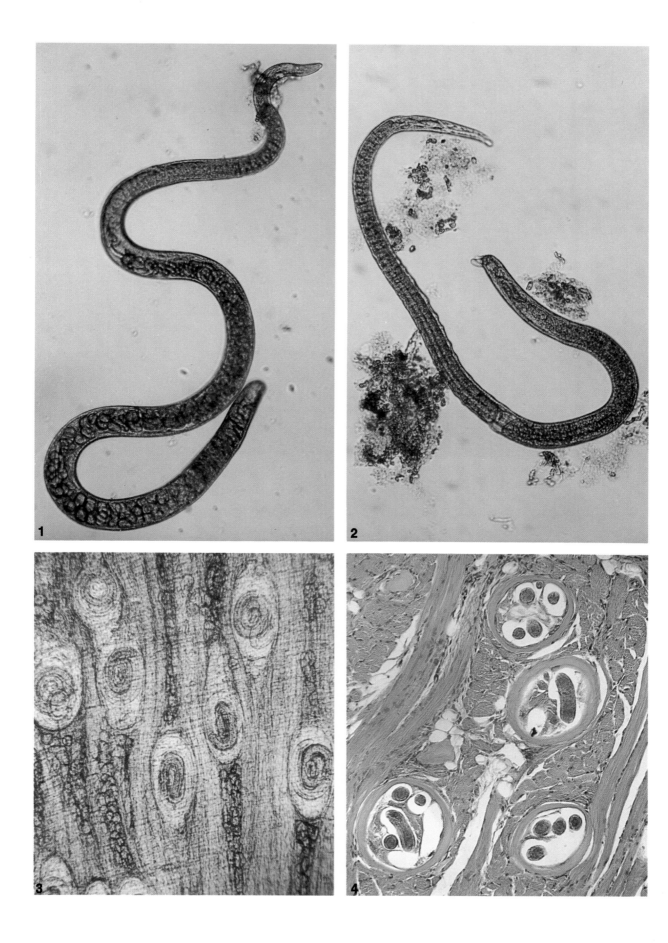

Uncommon Nematode Parasites in Humans

Many nematode parasites of animals occasionally parasitize humans. Frequently, the occurrence of these infections is regional and dependent upon the behavioral habits and occupations of particular populations.

Parasites of the genus *Physaloptera* live in the stomach and intestine of primates and a wide range of mammals, using various insects as intermediate hosts. The accidental ingestion of infected insects results in occasional human infections that are diagnosed by finding smooth, thick-shelled, broadly ovoid, embryonated eggs, measuring 44–65 μ long by 32–45 μ in feces.

Oesophagostomum species are bursate nematodes resembling hookworms and normally occur in monkeys and pigs. Human infection is especially common in Africa, but specific diagnosis and differentiation from human hookworm infections are difficult because the eggs are so similar in size and appearance.

Gongylonema, most important as a parasite of ruminants, may occasionally be found in the oral mucous and submucous membranes of humans. Various beetles and cockroaches serve as intermediate hosts. Eggs, which are similar to those of *Physaloptera*, are thick-shelled, measuring 50–70 μ by 25–37 μ, and embryonated when passed in feces. The lateral walls of the eggs usually are distinctly thickened.

Dioctophyma renale is the giant kidney worm of dogs, other canids, and various other wild mammals. Human infections have occurred when infected fish intermediate hosts are eaten uncooked. The unembryonated eggs are passed in urine, measure 60–80 μ by 39–46 μ, and have a thick, sculptured shell.

FIGURES. – Plate 39:1–4.

PLATE 39

Uncommon Nematode Parasites in Humans, Eggs

Fig 1. *Physaloptera* sp, egg in feces. This spiruroid parasite produces a small egg with a thick, hyaline shell that contains a larva when passed in feces. *P. caucasica* is a monkey and human parasite in Africa, with eggs that usually are 44–65 μ long by 32–45 μ wide.

Fig 2. *Oesophagostomum* sp, egg in feces. This egg closely resembles that of hookworms but is somewhat larger, and the germinal mass is in the advanced stage of division. The latter characteristically fills the egg at the time it is passed in feces. This parasite is seen most frequently in Africa.

Fig 3. *Gongylonema* sp, egg in feces. This egg has a thick, hyaline shell (much thicker on its sides), and, like other spiruroids, contains a larva when passed in feces. The eggs of *G. pulchrum*, a ruminant parasite that infects humans, typically measure 50–70 μ by 25–37 μ wide.

Fig 4. *Dioctophyma renale*, egg in urine. This unembryonated, ellipsoidal egg is found in the urine of infected animals. It is large, measuring 60–80 μ long by 39–46 μ wide. This egg has a thick shell with a wrinkled appearance on the sides, and the poles are a lighter color.

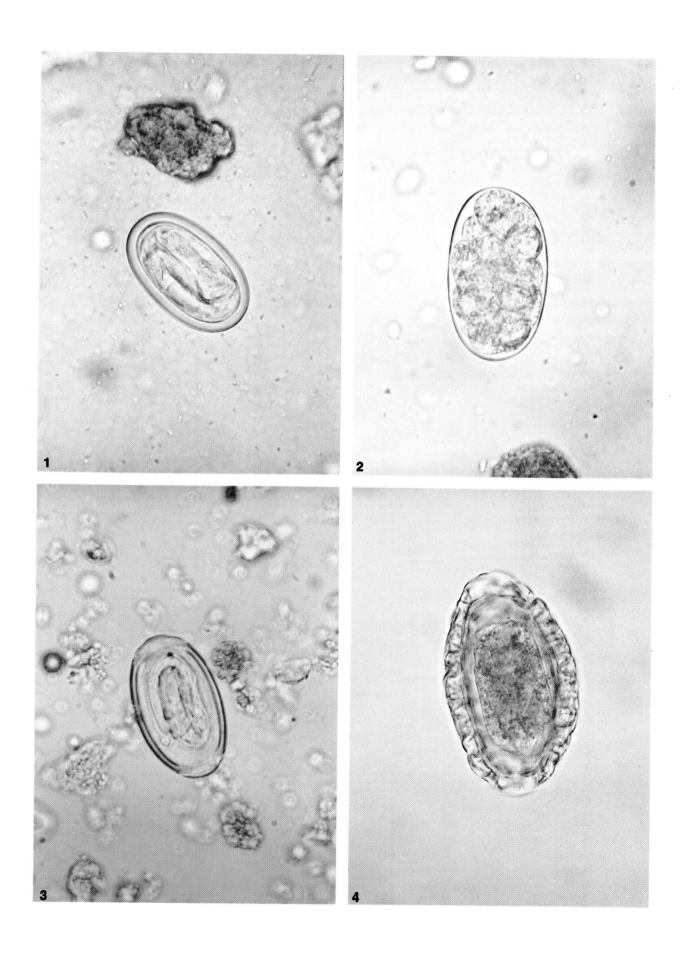

Toxocara canis

CLASSIFICATION. — Nematode.

DISEASE. — Visceral larva migrans.

GEOGRAPHIC DISTRIBUTION. — Cosmopolitan, but more common in warm, moist regions of the world.

LOCATION IN HOST. — Only the larval stage occurs in human tissues, especially the liver, eye, and central nervous system. In the normal host, ie, dogs and other canids, adult worms live in the small intestine.

MORPHOLOGY. — Adult worms. — Adult males are 4–6 cm long, and females are 6.5–10.0 cm long.

Eggs. — Eggs are thick-shelled and subspherical, with a markedly pitted surface resembling a golf ball. These measure 80–85 μ by 75 μ.

Larvae. — Second-stage larvae removed from human tissues measure 290–350 μ by 14–20 μ.

LIFE CYCLE. — Females in small intestine of dogs are oviparous. Unembryonated eggs pass in feces into soil, and development within the egg to the infective, second-stage larva takes two to three weeks. Young dogs may be infected by direct ingestion of infective eggs, by prenatal infection, or by ingestion of larval stages encapsulated in tissues of paratenic hosts. Humans, especially toddler-aged children, acquire infection by ingestion of infective eggs. Larvae become encapsulated in various organs and tissues of the body and live for long periods of time. This parasite occasionally has been reported to mature in humans, however, these cases lack good documentation.

DIAGNOSIS. — Canine infection is diagnosed by finding eggs in feces. Human infections usually are diagnosed based on clinical symptoms and identification of larvae in various tissues and organs. Serologic procedures may be useful.

Diagnostic problems. — Since humans are only infected with larval stages, establishment of diagnosis may be difficult.

COMMENTS. — T. cati, the cat ascarid, also has been identified as a cause of human visceral larva migrans. Eggs of this species are morphologically similar to those of T. canis, except that they are smaller (65–75 μ) and the surface of the shell is finely pitted.

Toxascaris leonina, an ascarid of dogs and cats, has not been incriminated as a cause of visceral larva migrans. The eggs of this species — though of similar size as T. canis and T. cati, measuring 75–85 μ by 60–75 μ — have a smooth shell and the internal vitelline membrane is prominently wrinkled in fresh specimens. The ovum usually is divided into dark and light granular hemispheres.

FIGURES. — Plate 40:1–6.

PLATE 40

Dog and Cat Ascarid Eggs

Figs 1 and 2. T. canis, unembryonated, fertile eggs in feces. This large egg, though never seen in human feces, is common in dogs, especially puppies. Its significance lies in the fact that ingestion of infective eggs from contaminated soil by humans may result in visceral larva migrans. The egg has a thick, pitted shell and is unembryonated when passed in feces. The characteristic surface pitting is best illustrated in Figure 2.

Figs 3 and 4. T. cati, unembryonated, fertile eggs in feces. The egg of this ascarid parasite of cats is somewhat smaller than that of T. canis, and the nature of the surface pitting is finer than that in T. canis. Compare Figures 2 and 4. This parasite also may cause human visceral larva migrans.

Figs 5 and 6. Toxascaris leonina, unembryonated, fertile eggs in feces. This ascarid parasite occurs both in cats and dogs, but has not been incriminated as a cause of human visceral larva migrans. The egg can be differentiated from T. canis and T. cati by its smooth, thick shell, and the prominent, internal, wrinkled vitelline membrane. Also, in fresh fecal specimens, the ovum usually is seen to be distinctly divided into light and dark hemispheres.

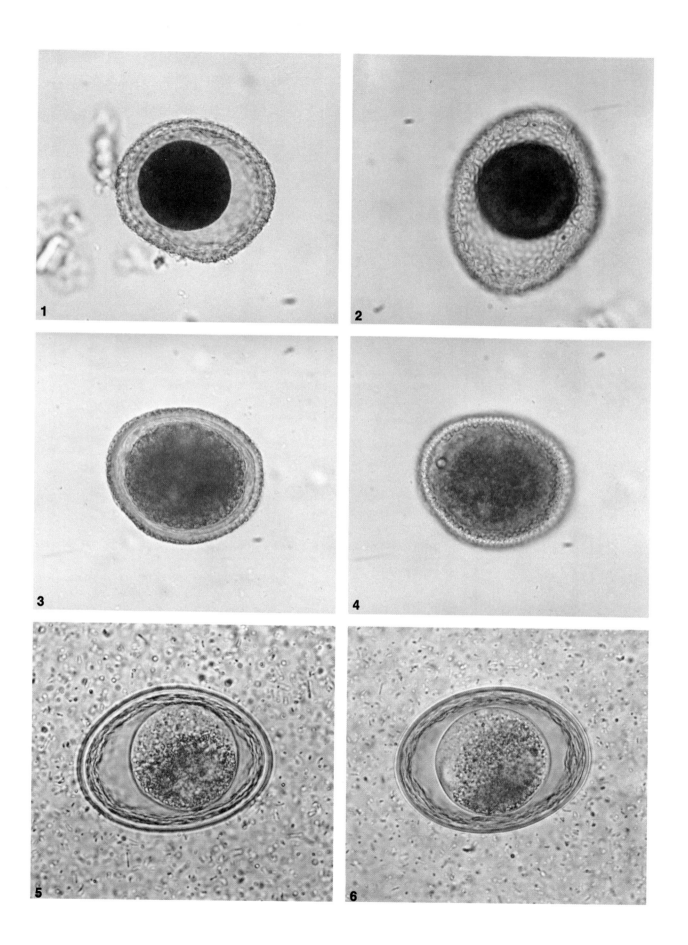

PLATE 41

Fecal Elements

Fig 1. Plant hairs—in this example, peach fuzz—frequently are confused with *Strongyloides* or other nematode larvae by inexperienced workers. The refractile, amorphous central canal, which runs the length of the hair, should not be taken for the well differentiated esophagus and intestine of a nematode larva. Although one end of the hair may be tapering and smoothly rounded, the other end usually is blunt and irregular. This feature, of course, is not seen in nematode larvae.

Fig 2. Vegetable cell. Such cells sometimes are confused with helminth eggs. However, the irregular contour of the cell wall and the internal contents should clearly indicate this to be plant material. When stained with iodine, these cells usually appear dark brown.

Fig 3. Charcot-Leyden crystals are breakdown products of eosinophils and frequently are found in sputum or feces. They are six-sided crystals often occurring in persons with parasitic or other infections eliciting an immune response.

Fig 4. Vegetable spiral. Part of the spirovascular bundle of plants, such structures may be loosely or tightly coiled and of varying size.

1

2

3

4

PLATE 42

Fecal Elements

Figs 1 and 2. These are representative of plant materials that inexperienced microscopists sometimes confuse with parasitic elements.

Figs 3, 5, and 6. Pollen grains frequently are seen in feces and may be confused with parasites. The presence of a striated outer wall may lead to confusion with *Taenia* eggs. However, the small size, the uneven thickness of the wall, and the nature of the internal contents should indicate that this is not an egg.

Fig 4. Diatoms occur in water. Their presence in fecal smears usually is indicative of contamination of solutions used in preparing the smears.

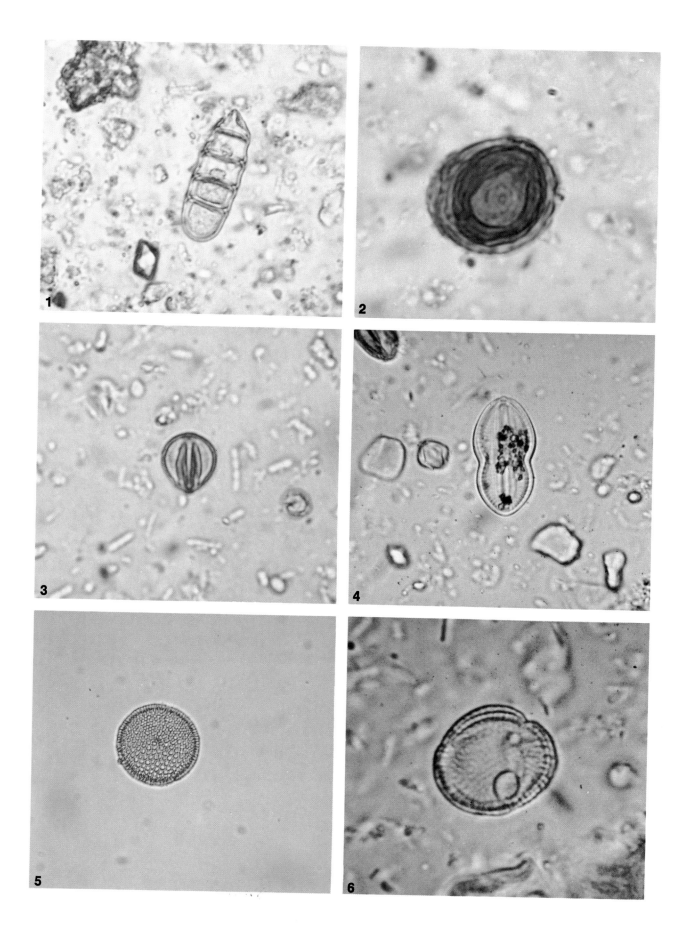

PLATE 43

Fecal Elements

Fig 1. Plant material. This structure has a thick wall and might be confused with a helminth egg. However, the wall is of uneven thickness and the object has an irregular shape.

Fig 2. This mite egg occasionally is found in feces and may be confused with hookworm or trichostrongyle eggs. However, the former's larger size should aid in differentiation, as mite eggs usually are more than 100 μ long.

Fig 3. *Heterodera (Meloidogyne)* egg. This is the egg of a plant parasitic nematode that occurs on root vegetables such as radishes and turnips. The eggs usually are between 80–120 μ by 25–40 μ, and can be confused with hookworm eggs. The larvae of these nematodes also may be found occasionally in feces.

Fig 4. This structure, known as a "Beaver body," is *Psorospermium haeckelii*, a stage of an alga that occurs in the tissues of crayfish. It sometimes is confused with helminth eggs when found in the feces of persons who have eaten crayfish.

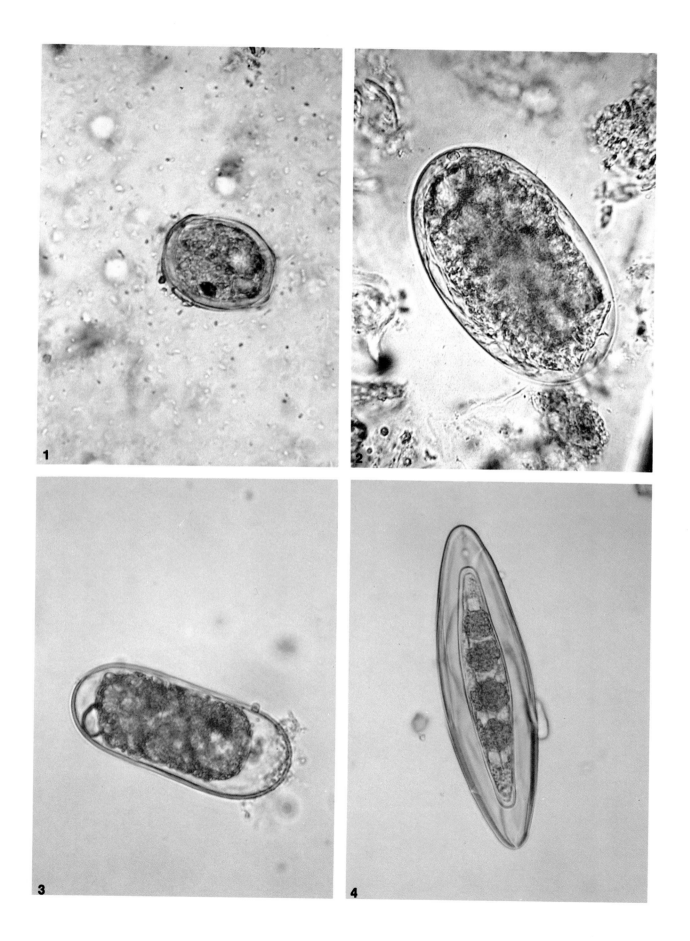

1

2

3

4

Wuchereria bancrofti

CLASSIFICATION. — Nematode.

DISEASE. — Bancroft's filariasis, elephantiasis.

GEOGRAPHIC DISTRIBUTION. — Widely distributed throughout tropics and subtropics, Indian subcontinent, most of Africa except South Africa, China, Japan, and southeast Asia. In Europe, distribution is limited to Turkey, southern Italy, Yugoslavia, and Hungary. In the Americas, this parasite occurs in Costa Rica, Guianas, Brazil, and islands of the West Indies. The organism also occurs on many islands in the South Pacific: Micronesia, New Guinea, and the Solomons. The subperiodic form of the parasite occurs in New Caledonia, Fiji, Tonga, Western and American Samoa, Cook Islands, and islands of French Polynesia.

LOCATION IN HOST. — Adult worms live chiefly in lymphatic vessels but occasionally may be found in blood vessels. Microfilariae circulate in blood.

MORPHOLOGY. — Adult worms. — These worms are thread-like; males measure up to 40 mm long by 0.1 mm in diameter, and females are 80 – 100 mm by 0.25 mm.

Microfilariae. — Microfilariae are sheathed. In stained blood films, they measure 244 – 296 μ (mean, 260 μ) by 7.5 – 10.0 μ. In 2% formalin, they measure 275 – 317 μ (mean, 298 μ) by 7.5 – 10.0 μ. The tail tapers to a point and nuclei in the tail stop short of the end of the tail.

LIFE CYCLE. — Females produce sheathed microfilariae, which, in most areas of the world except the South Pacific, circulate in the peripheral blood during the evening hours (nocturnal periodicity) and are ingested by mosquitoes of the genera Culex, Anopheles, Aedes, and Mansonia. Development to the infective third stage in the mosquito vector requires approximately two weeks. Infection is transmitted to others by bites of infected mosquitoes.

Although the time required for development to the adult stage is not well established, it apparently ranges from four months to a year. Adult worms usually live five to ten years but may survive for even longer periods. Microfilariae probably live for one to two years.

DIAGNOSIS. — Demonstration of sheathed microfilariae in blood. With Giemsa's stain, the sheath stains poorly or not at all; an Innenkörper (inner body) is present.

Diagnostic problems. — For the most common strain of this parasite, ie, the nocturnally periodic form, it is essential to examine the blood between 9 PM and 3 AM. Microfilariae are absent from the blood and accumulate in the lungs and other visceral organs during other hours of the day. The subperiodic form of the parasite may have microfilariae in the blood at all times of the day, with some peaking in the evening hours. Microfilariae may be present in small numbers in the blood, and concentration procedures — thick blood films, Knott concentration, or millipore concentration procedures — may be used to detect infections. Microfilariae must be distinguished from other sheathed microfilariae. This generally is done most easily on the basis of the arrangement of nuclei in the tail.

FIGURES. — Plate 44:1 – 4; Plate 45:1 – 4.

PLATE 44

Wuchereria bancrofti, Microfilariae, Hematoxylin Stain

Figs 1 and 2. In hematoxylin-stained thick blood films, it is evident that the microfilaria has a sheath, short cephalic space, well demarcated anatomic landmarks, and a pointed tail devoid of nuclei. Occasionally, the lightly staining sheath is shed by the microfilaria before the blood film dries (Fig 2).

Figs 3 and 4. At higher magnification, the discrete nature of the nuclei in the nuclear column is well demonstrated in thick blood films. The shape of the tail and the distribution of nuclei therein are especially clear in Figure 4.

114

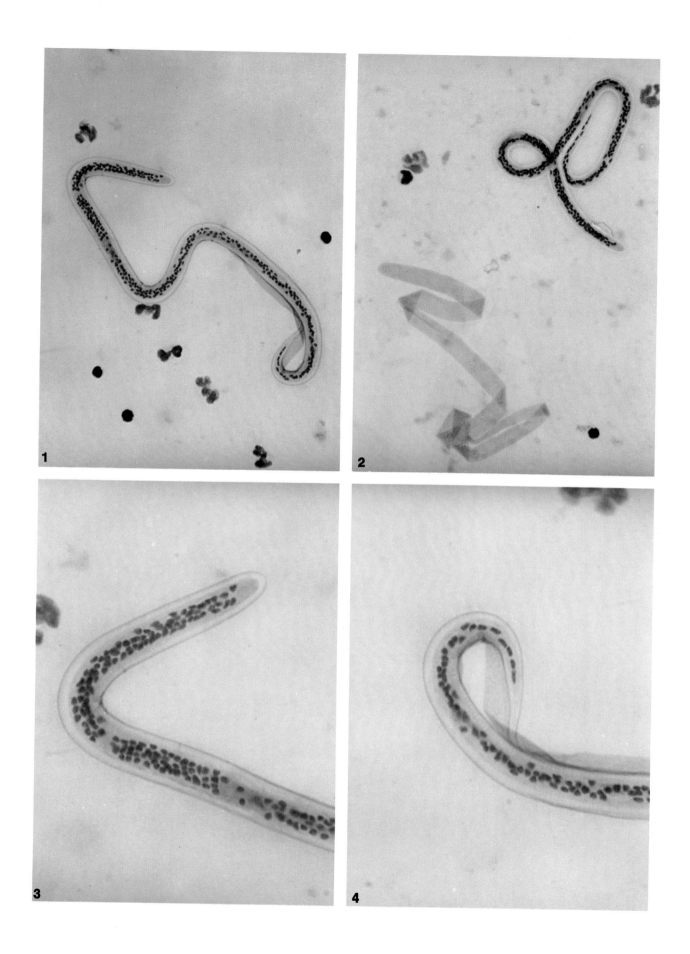

PLATE 45

Wuchereria bancrofti, Microfilariae in Thick Blood Films

Figs 1 and 2. Microfilariae stained with hematoxylin. The microfilariae in thick blood films usually are in graceful curves, and the sheaths may be seen projecting from either end, with a large portion often trailing from the posterior end.

Figs 3 and 4. Microfilariae in thick films, Giemsa's stain. In thick or thin films stained by this method, the sheath usually fails to stain and may appear only as a halo or clear space along the margins of the microfilaria. The Innenkörper (inner body), located in the posterior half of the body (Fig 3, *arrow*), takes a reddish-pink stain. This is further illustrated in Figure 4 at higher magnification.

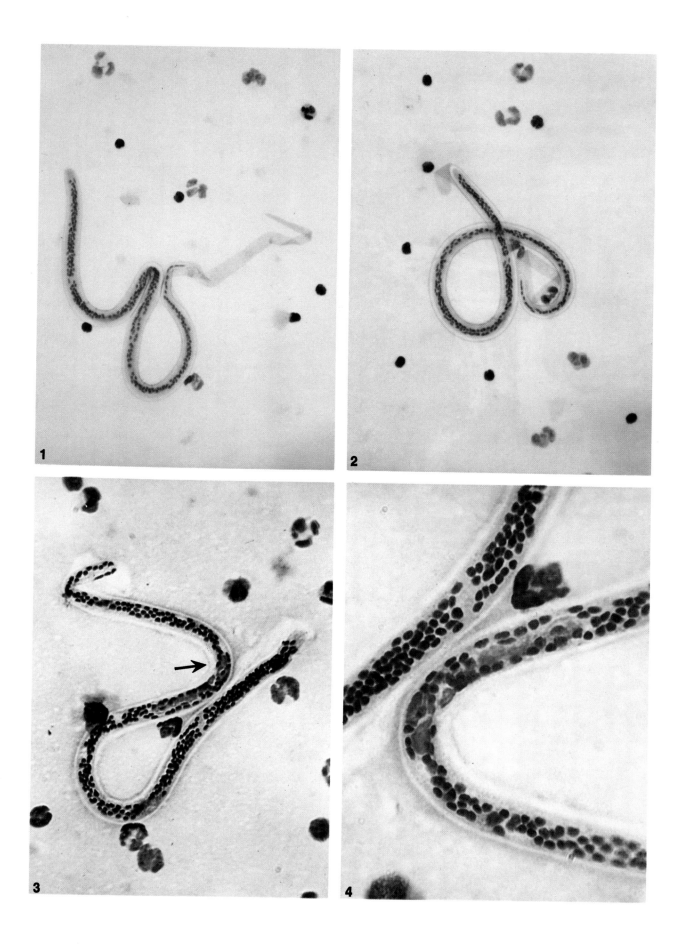

Brugia malayi

CLASSIFICATION.—Nematode.

DISEASE.—Malayan filariasis, elephantiasis.

GEOGRAPHIC DISTRIBUTION.—Southeast Asia, Korea, China, India, and Sri Lanka.

LOCATION IN HOST.—Adult worms live mainly in lymphatic vessels, but occasionally may occur in blood vessels. Microfilariae circulate in blood.

MORPHOLOGY.—Adult worms.—Adults are thread-like; males measure 13–25 mm by 70–80 μ, and females measure 43–55 mm by 130–170 μ.

Microfilariae.—Microfilariae are sheathed. In stained blood films, they measure 177–230 μ (mean, 220 μ) by 5–6 μ. In 2% formalin, microfilariae measure 240–298 μ (mean, 270 μ) by 5–6 μ. The tail is tapered, and a constriction separates the column of nuclei from the last nucleus at the end of the tail.

LIFE CYCLE.—Females produce sheathed microfilariae that typically are found in the bloodstream during evening hours (nocturnal periodicity). Microfilariae will develop into the infective stage in mosquitoes of the genera *Mansonia, Aedes,* and *Anopheles.* Development to the infective stage takes approximately 10–14 days and transmission of infection is by bite of infected mosquitoes. In the human host, development to maturity takes approximately three to five months.

Whereas the periodic form of the parasite essentially is a human disease without important animal reservoirs, a subperiodic form of the parasite associated with swamplands also exists.

DIAGNOSIS.—Demonstration of sheathed microfilariae in blood. The sheath stains pink with Giemsa's stain. There is a conspicuous Innenkörper. The characteristic arrangement of nuclei in the tail serves to separate this species from *Wuchereria bancrofti.*

Diagnostic problems.—This species may be confused with *W. bancrofti* or *B. timori;* however, distribution of tail nuclei and staining properties of the sheath are diagnostic aids. For further differentiation from *B. timori* see the description preceding Plate 52.

FIGURES.—Plate 46: 1–4; Plate 47: 1–4.

PLATE 46

Brugia malayi, Microfilariae, Hematoxylin Stain

Fig 1. This microfilaria is about the same size as that of *W. bancrofti* and has a sheath. In contrast to *W. bancrofti,* the *B. malayi* microfilaria has a long cephalic space, a compact column of nuclei, and a tail with a subterminal and a terminal nucleus separated by a short constriction. In thick films, this microfilaria does not exhibit the smooth, graceful curves of *W. bancrofti.*

Fig 2. Microfilaria from Knott concentration. This microfilaria demonstrates the typical morphologic features of this species, following preservation and concentration in 2% formalin, and staining with hematoxylin. Note the dot-like terminal and subterminal nuclei in the tail.

Figs 3 and 4. Microfilariae in thick blood films. The terminal and subterminal nuclei may not be evident in all microfilariae. Note the apparent absence of the subterminal nucleus in Figure 3. At high magnification (Fig 4), all the salient morphologic features are evident.

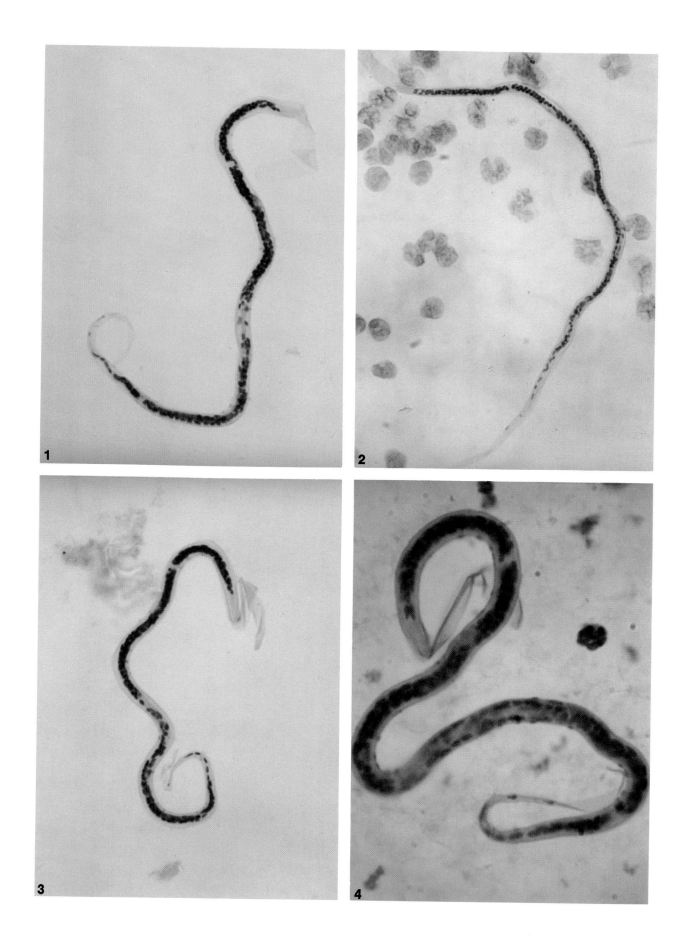

PLATE 47

Brugia malayi, Microfilariae, Thick Blood Films, Giemsa's Stain

Figs 1 and 2. In Giemsa-stained thick or thin blood films, the sheath of the microfilaria takes a pink stain, and the nuclear column stains an intense blue so that individual nuclei may be difficult to see.

Figs 3 and 4. At high magnification, note the long cephalic space and the structure of the tail. Even though the terminal and subterminal nuclei *(arrows)* have failed to stain, their position in the tail and the constriction between them is apparent in Figure 4.

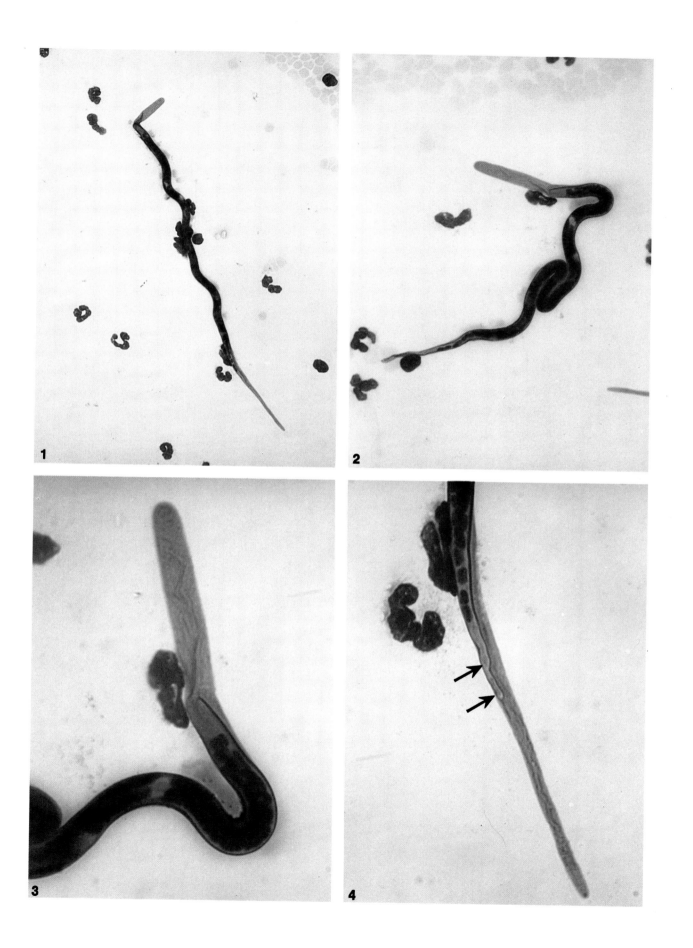

Loa loa

CLASSIFICATION. — Nematode.

DISEASE. — Loiasis, African eye worm.

GEOGRAPHIC DISTRIBUTION. — Equatorial rain forest regions of west and central Africa, south of the Sahara desert.

LOCATION IN HOST. — Adult worms occur in subcutaneous tissues; microfilariae circulate in blood.

MORPHOLOGY. — Adult worms. — Thread-like, with adult males measuring 30–35 mm by 0.35–0.43 mm, and females measuring 50–70 mm by 0.5 mm.

Microfilariae. — In stained blood films, these sheathed microfilariae measure 231–250 μ (mean, 238 μ) by 6.0–8.5 μ. Microfilariae in formalin measure 270–300 μ (mean, 281 μ) by 6.0–8.5 μ. The tail is tapered, with nuclei extending to the tip of the tail.

LIFE CYCLE. — Microfilariae in peripheral blood are taken up by bloodsucking tabanid flies of the genus *Chrysops*. Development to the infective stage in the fly takes 10–12 days, and transmission to humans is by the bites of the infected flies. The prepatent period in humans is about five to six months. This infection may be long-lived, lasting for several years or longer.

DIAGNOSIS. — Presence of characteristic microfilariae in blood. Microfilariae are found in blood only during daylight hours (diurnal periodicity). The sheath does not stain with Giemsa's stain.

Diagnostic problems. — The microfilariae must be distinguished from other sheathed microfilariae of the genera *Wuchereria* and *Brugia*. However, the presence of nuclei throughout the length of the tail should clearly distinguish the species. Furthermore, the diurnal periodicity of this microfilaria differentiates it from *Wuchereria* and *Brugia*.

COMMENTS. — Adult worms have a propensity for wandering through the superficial layers of the subcutaneous tissues, producing temporary, transient, inflammatory reactions usually referred to as "Calabar" swellings. Frequently, the worms enter the orbit and move through the conjunctivae. These worms often are removed by the physician and sent to the laboratory for identification.

FIGURES. — Plate 48: 1–4; Plate 49: 1–4.

PLATE 48

Loa loa, Microfilariae in Thick Blood Films, Hematoxylin Stain

Figs 1–3. The microfilaria of this species is large and has a sheath. The nuclear column is densely stained and compact, and, distally, it is reduced to a column of five or six single, unevenly spaced nuclei. Occasionally, the microfilaria may shed its sheath *(arrow)* in thick blood films (Fig 3).

Fig 4. This microfilaria, the largest of all the species found in humans, sometimes is found as a mixed infection with other species. In these cases, the microfilaria's size may be a useful criterion for its identification. In this figure, it is seen in combination with the smaller, unsheathed microfilaria of *Dipetalonema perstans*.

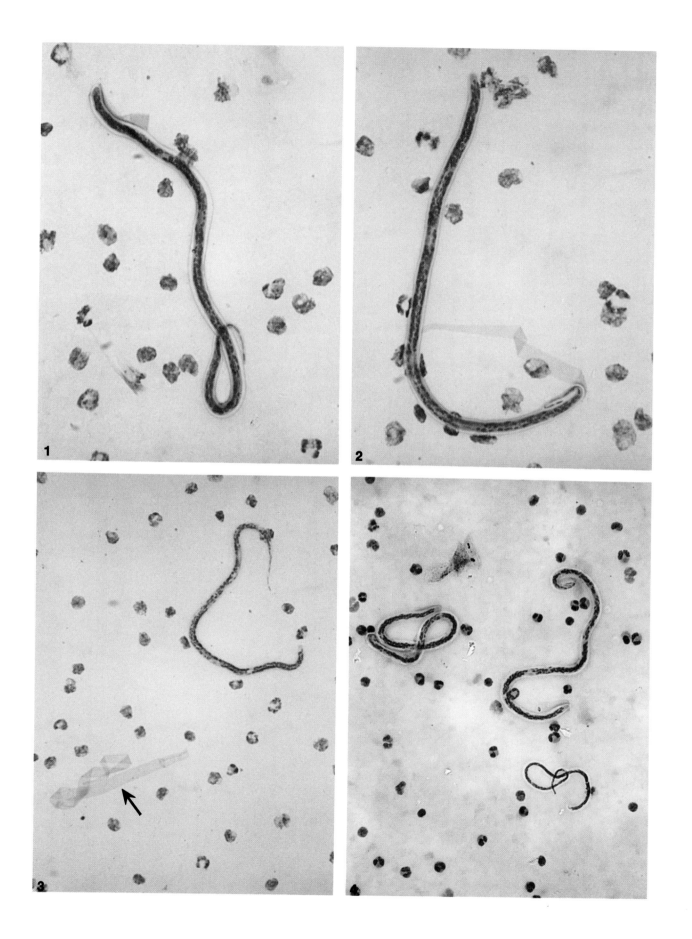

PLATE 49

Loa loa, Microfilariae, Giemsa's Stain

Figs 1 and 2. The sheath of the microfilaria does not stain in thin or thick Giemsa-stained blood films. However, its presence is marked by a clear, halo-like area around the body of the microfilaria in thick films, or by an outline of red blood cells lying adjacent to it in thin films (Fig 1). The short cephalic space, nerve ring, and other anatomic landmarks are usually evident.

Figs 3 and 4. Even though the sheath is not discernible in Giemsa-stained preparations, at higher magnifications the distribution and spacing of the last five or six nuclei in the tail are sufficient to identify the species. The terminal nucleus invariably lies near the tip of the tail.

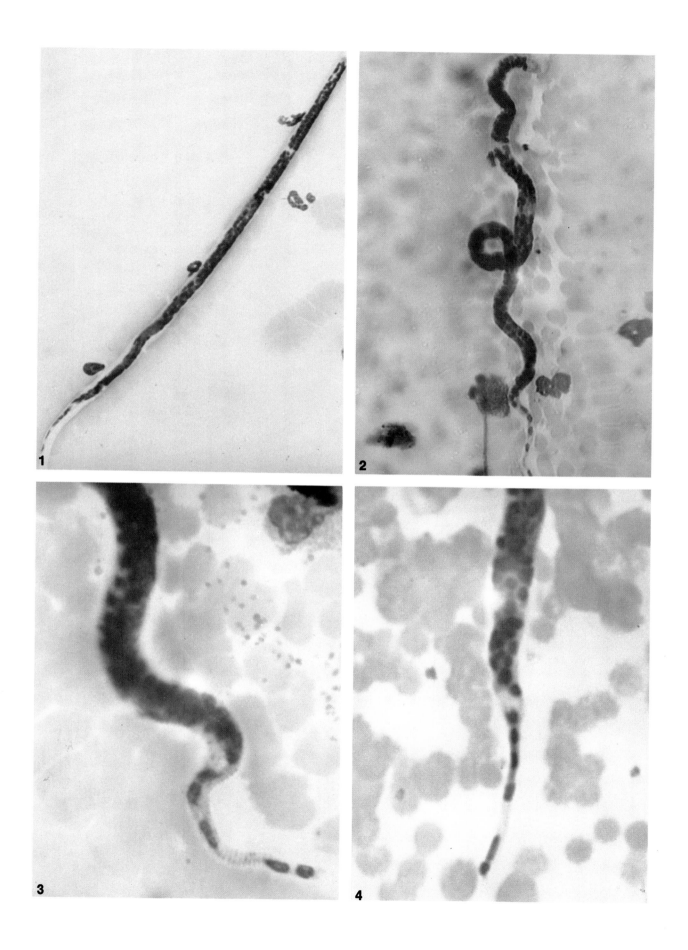

Dipetalonema perstans

CLASSIFICATION. — Nematode.

DISEASE. — Filariasis perstans, dipetalonemiasis.

GEOGRAPHIC DISTRIBUTION. — West and central Africa, and parts of east Africa south of the Sahara Desert; many parts of South America and certain Caribbean Islands.

LOCATION IN HOST. — Adult worms inhabit body cavities, mesenteries, and perirenal tissues of humans; microfilariae circulate in blood.

MORPHOLOGY. — Adult worms. — Adult males are approximately 45 mm by 60 μ, females are 70–80 mm by 120 μ.

Microfilariae. — In blood films, the unsheathed microfilariae measure 190–200 μ (mean, 195 μ) by 4 μ. In 2% formalin, they measure 183–225 μ (mean, 203 μ) by 4 μ. The tail tapers to a bluntly rounded end and nuclei extend to the end of the tail.

LIFE CYCLE. — Microfilariae circulating in blood are picked up by biting midges (*Culicoides* spp). Development to the infective stage takes seven to ten days. The prepatent period is believed to be about three to five months.

DIAGNOSIS. — Demonstration of characteristic microfilariae in blood.

Diagnostic problems. — Microfilariae are small and sometimes can be overlooked in blood films.

FIGURES. — Plate 50:1–4.

PLATE 50

Dipetalonema perstans, Microfilariae in Thick Blood Films

Figs 1 and 2. Thick blood films stained with hematoxylin. This small, unsheathed microfilaria has a densely stained column of nuclei extending to the bluntly rounded, posterior end. Landmarks such as the cephalic space and excretory pore are not always clearly demarcated.

Fig 3. Thick blood film, Giemsa's stain. The morphologic features of this microfilaria when stained with Giemsa's stain are essentially identical to those stained with hematoxylin. The blunt tail is a key characteristic in the identification of this species.

Fig 4. Thick blood film, hematoxylin stain. In the western hemisphere, *D. perstans* frequently occurs in combination with another unsheathed filarial species, *Mansonella ozzardi*. The two shown here, though of comparable size, can be distinguished on the basis of the slender, pointed tail of *M. ozzardi* (lower right).

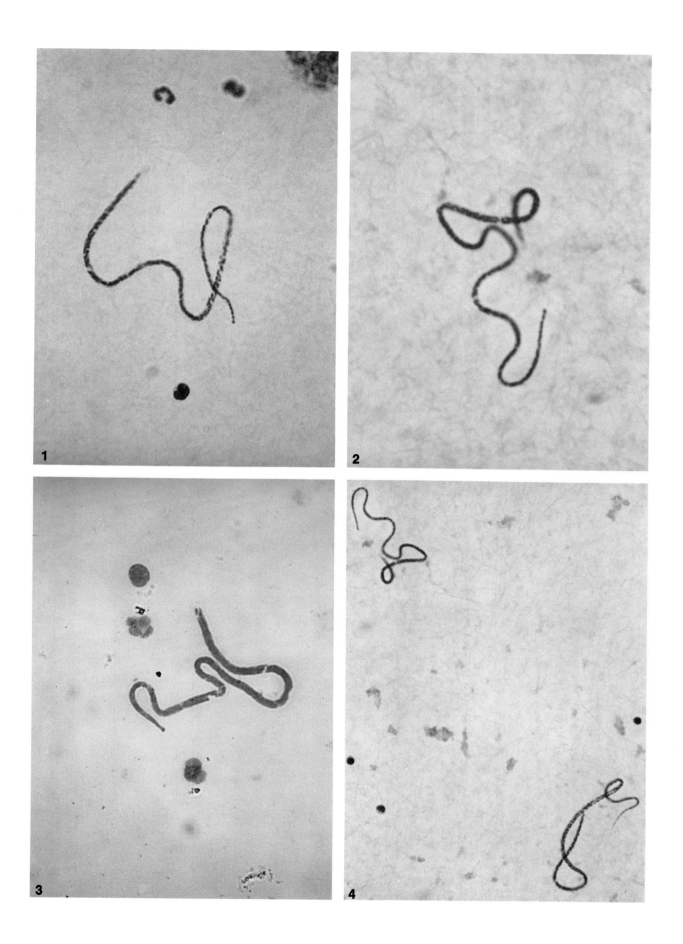

Mansonella ozzardi

CLASSIFICATION. — Nematode.

DISEASE. — Mansonellosis.

GEOGRAPHIC DISTRIBUTION. — South and Central America, Mexico, and the West Indies.

LOCATION IN HOST. — Adult worms are presumed to inhabit body cavities, mesenteries, and visceral fat. Microfilariae are found in blood, but reports indicate that they occasionally may be encountered in bloodless skin snips from infected persons.

MORPHOLOGY. — Adult worms. — These worms have been collected only on rare occasions and consequently are poorly described. Males are approximately 40 mm long by 0.2 mm wide; females, 65–81 mm by 0.2–0.25 mm.

Microfilariae. — The small, unsheathed microfilariae are 163–203 μ (mean, 183 μ) by 3–4 μ in stained blood films. In 2% formalin, they measure 203–254 μ (mean, 224 μ) by 4–5 μ. Microfilariae have a long, slender tail, with the nuclear column terminating considerably short of the end of the tail. The latter is characteristically bent in buttonhook fashion in formalin-preserved material.

LIFE CYCLE. — Unsheathed microfilariae are taken up by arthropod vectors. In the Caribbean, biting midges of the genus *Culicoides* apparently are the intermediate host, but in Brazil and Colombia, black flies (*Simulium*) appear to be the vector. Very little is known about the vertebrate phase of the life cycle. The prepatent period of development is speculated to be six months or less.

DIAGNOSIS. — Demonstration of microfilariae in blood. Microfilariae exhibit no periodicity.

Diagnostic problems. — Microfilariae are small and can sometimes be overlooked in blood films. Because of their size, they may be confused with *Dipetalonema perstans*.

FIGURES. — Plate 51: 1–4.

PLATE 51

Mansonella ozzardi, Microfilariae in Thick Blood Films

Figs 1 and 2. Thick blood films, hematoxylin stain. The cells of the nuclear column are more or less discrete, depending on the type of fixative used. In any case, the microfilaria may be identified by its small size, absence of a sheath, and the slender, attenuated tail.

Figs 3 and 4. Giemsa-stained thick blood films. The microfilaria is small, lacks a sheath, has a densely stained nuclear column, and a long, slender pointed tail. The column of nuclei does not extend to the end of the tail. However, that portion of the tail devoid of nuclei usually is difficult to see in most preparations.

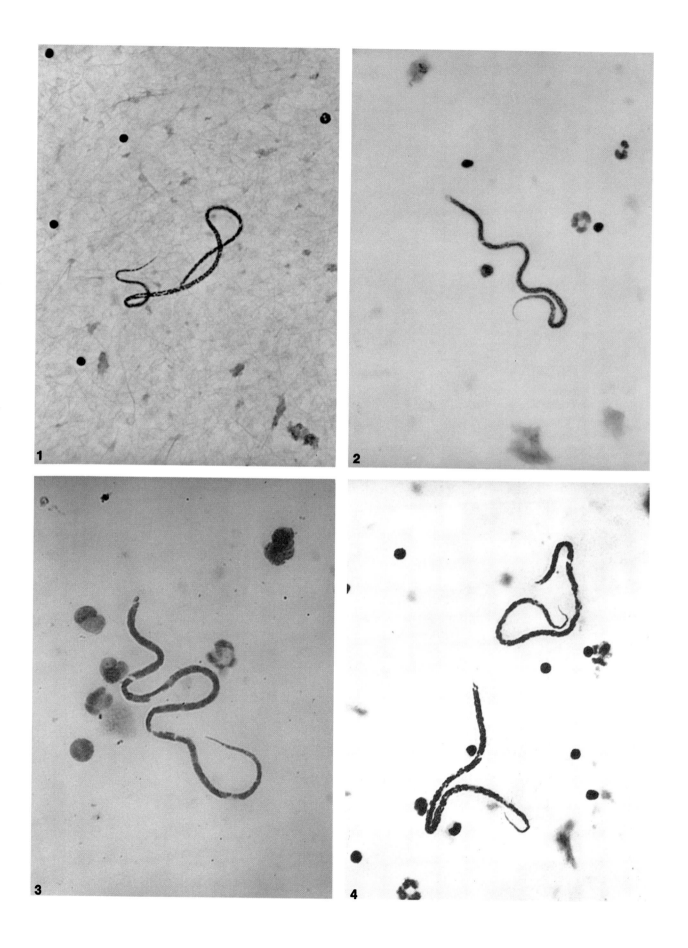

Other Human Filariae: *Brugia timori* and *Dipetalonema semiclarum*

CLASSIFICATION. — Nematodes.

DISEASES. — Timor filariasis *(B. timori)* and dipetalonemiasis *(D. semiclarum)*.

GEOGRAPHIC DISTRIBUTION. — Lesser Sunda Islands of Indonesian archipelago *(B. timori)* and the central African country of Zaire *(D. semiclarum)*.

LOCATION IN HOST. — *B. timori* adult worms are found in lymphatic vessels, and microfilariae are found in blood. *D. semiclarum* microfilariae occur in blood and skin, and adult worms are unknown.

MORPHOLOGY. — Adult worms. — The adult worms of *B. timori* are known only from experimental rodent infections. Male worms have an average length of 17 mm and a diameter of 0.07 mm. Females measure 26 × 0.10 mm. The adult worms of *D. semiclarum* have not been described.

Microfilariae. — In *B. timori*, microfilariae are sheathed and have mean lengths of 310 μ by 6–7 μ (in blood films), and 341 μ by 6–7 μ (in formalin). The tail does not have a prominent constriction near its end, and there is some distance between the subterminal and the terminal nucleus. The microfilariae of *D. semiclarum* are longer and broader (mean, 198 μ by 5.2 μ in blood films) than those of *D. perstans*. *D. semiclarum* has a characteristic clear band measuring 25–40 μ long in the posterior half of the body.

DIAGNOSIS. — Demonstration of microfilariae in blood.

Diagnostic problems. — In the areas where *B. timori* is found, the microfilaria must be differentiated from *B. malayi* and *W. bancrofti*. *B. timori* is differentiated from *B. malayi* by a sheath that does not stain with Giemsa's stain, a longer cephalic space, greater numbers of single-row nuclei toward the posterior end, and a lesser bulge of the cuticle around the subterminal nucleus.

D. semiclarum and *D. perstans* may occur in the same geographic area. Although both species are of similar size and are unsheathed, the clear area in the posterior half of the body should identify *D. semiclarum*.

FIGURES. — Plate 52: 1–4.

PLATE 52

Brugia timori and *Dipetalonema semiclarum*, Microfilariae

Figs 1 and 2. *B. timori*, microfilariae in thick blood films, Giemsa's stain. This microfilaria, first described in humans in 1967, has the characteristics of the genus *Brugia*. The microfilaria has a sheath, a densely stained nuclear column, and the characteristic terminal and subterminal tail nuclei. However, as illustrated here, and in contrast to *B. malayi*, the sheath does not stain with Giemsa's stain.

Figs 3 and 4. *D. semiclarum*, microfilarie in thick blood films, hematoxylin stain. In 1974, this microfilaria was described in people from the central African country of Zaire. Although this microfilaria bears a superficial resemblance to *D. perstans*, which occurs in the same geographic area, *D. semiclarum* is longer and broader, has a different shape to the anterior end, and a shorter head space. In Figure 4, both *D. semiclarum* and *D. perstans* are seen in the same field. *D. semiclarum* (near the left margin) is larger and displays the anucleate area typical of this species.

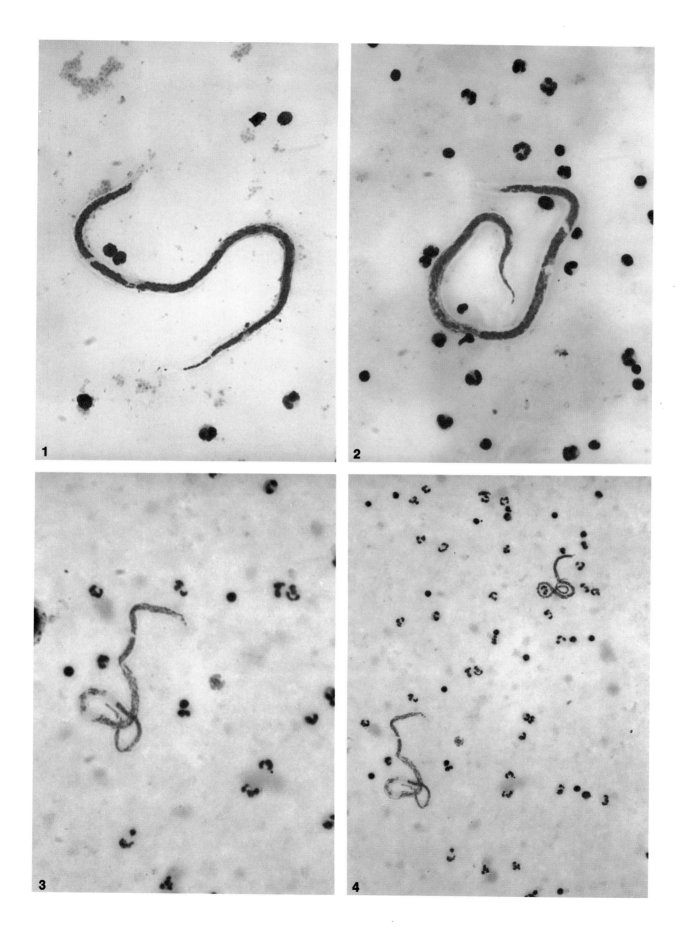

Onchocerca volvulus

CLASSIFICATION. — Nematode.

DISEASE. — Onchocerciasis, river blindness.

GEOGRAPHIC DISTRIBUTION. — West and central Africa, south of the Sahara Desert; southern Mexico, Central America, and northern South America.

LOCATION IN HOST. — Adult worms are typically found in nodules in subcutaneous tissues; microfilariae occur in skin.

MORPHOLOGY. — Adult worms. — Adult males are 19–42 mm by 0.13–0.21 mm; females are 33–50 cm by 0.27–0.40 mm.

Microfilariae. — The unsheathed microfilaria measures 304–315 μ (mean, 309 μ) by 5–9 μ. The tail tapers to a point and often is sharply flexed. The column of nuclei does not extend to the end of the tail.

LIFE CYCLE. — Microfilariae in skin are taken up in tissue juices when black flies (genus *Simu-lium*) bite. Development to the infective stage takes approximately seven days, and infection is transmitted by the bite of infected black flies. It takes approximately 9–12 months for adult worms to mature and for microfilariae to appear in the skin. Adult worms may live for 10–15 years.

DIAGNOSIS. — Demonstration of characteristic, unsheathed microfilariae in skin snips. Microfilariae may appear in peripheral blood and urine following administration of the drug, diethylcarbamazine.

Diagnostic problems. — In Africa, where this infection coexists with *Dipetalonema streptocerca*, *O. volvulus* microfilariae in skin must be distinguished from those of *D. streptocerca*, which also occur in skin. The two microfilariae are distinguished from each other on the basis of size and characteristics of the nuclear column and the tail.

FIGURES. — Plate 53:1–4.

PLATE 53

Onchocerca volvulus, Microfilariae

Fig 1. Unstained microfilaria teased from skin snip. The unstained microfilaria is large, lacks a sheath, and has an attenuated, pointed tail that invariably is flexed.

Fig 2. Giemsa-stained microfilaria from teased skin snip. In Giemsa-stained preparations, the microfilaria displays the morphologic features described in Figure 1. In addition, the nuclear column may be densely packed and deeply stained. The cephalic space is longer than it is wide, and the nerve ring space is well demarcated. The nuclear column does not extend to the end of the body. The terminal portion of the tail, which lacks nuclei, may be difficult to see in some preparations.

Fig 3. Hematoxylin-stained microfilaria teased from skin snip. All of the morphologic features described in Figures 1 and 2 are evident in specimens stained with hematoxylin.

Fig 4. In this hematoxylin-eosin-stained section of human skin, one can see portions of microfilariae (*arrows*) in the superficial layer of the dermis.

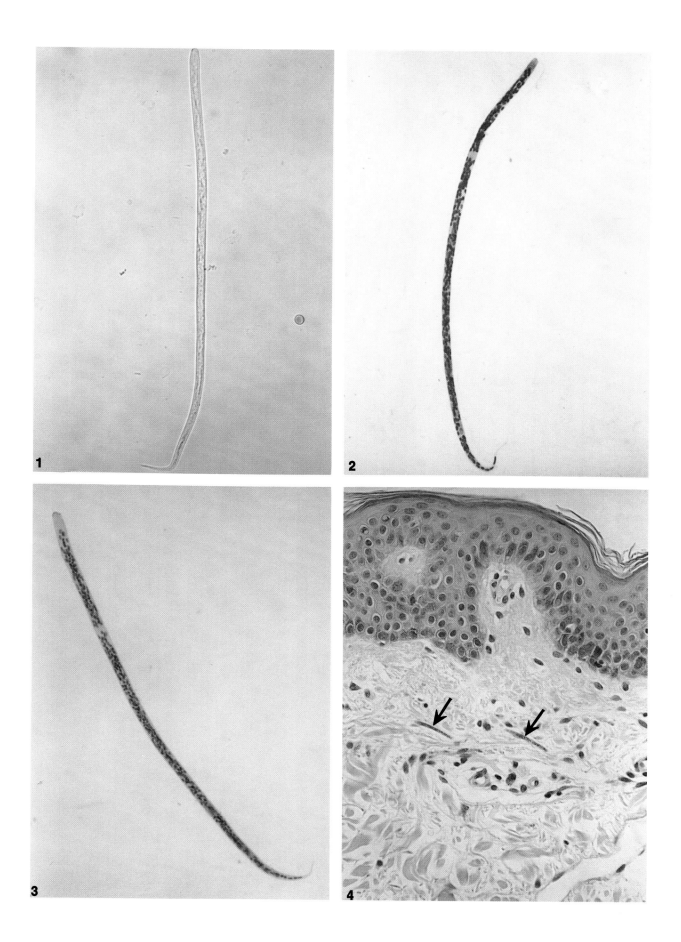

Dipetalonema streptocerca

CLASSIFICATION. — Nematode.

DISEASE. — Dipetalonemiasis, streptocerciasis.

GEOGRAPHIC DISTRIBUTION. — Central African rain forest areas including Zaire, Ghana, Nigeria, and Cameroons.

LOCATION IN HOST. — Adult worms occur in dermal and subcutaneous tissue. Microfilariae do not circulate in blood, but are found in skin.

MORPHOLOGY. — Adult worms. — Males are approximately 13–17 mm long by 47 μ wide. Females are 19–25 mm long by 76 μ wide.

Microfilariae. — Microfilariae are unsheathed, measuring 180–240 μ (mean, 210 μ) long by 5–6 μ wide. The tail is characteristically bent into a buttonhook shape, with nuclei extending to the end of the body.

LIFE CYCLE. — Microfilariae in skin are picked up by biting midges (*Culicoides* spp) that serve as the intermediate host. Transmission of infection to humans is by bite of infected culicoids. The time required to reach maturation, although not definitely known, is thought to be similar to that observed for *D. perstans*, ie, from a few to several months.

DIAGNOSIS. — By demonstration in skin snips of microfilariae with characteristic morphologic features.

Diagnostic problems. — In areas of Africa where this species and *Onchocerca volvulus* coexist, *D. streptocerca* must be distinguished from the much larger skin-dwelling microfilariae of *O. volvulus*. In light infections, microfilariae in skin may be difficult to detect unless several skin snips are examined.

FIGURES. — Plate 54:1–4; Plate 55:1–3.

PLATE 54

Dipetalonema streptocerca, Microfilariae

Figs 1 and 2. Hematoxylin-stained microfilariae teased from skin. This species of microfilaria is of moderate size, long and slender, lacks a sheath, and has a nuclear column extending to the end of the body. The tail is characteristically bent into the shape of a shepherd's crook. The body of the microfilaria usually is straight.

Figs 3 and 4. Anterior and posterior ends of this species at higher magnification. In Figure 3, note that the nuclear column begins as a single row of nine or ten nuclei. In the posterior end of the microfilaria, illustrated in Figure 4, the nuclear column is reduced to a single row that extends to the tip of the tail.

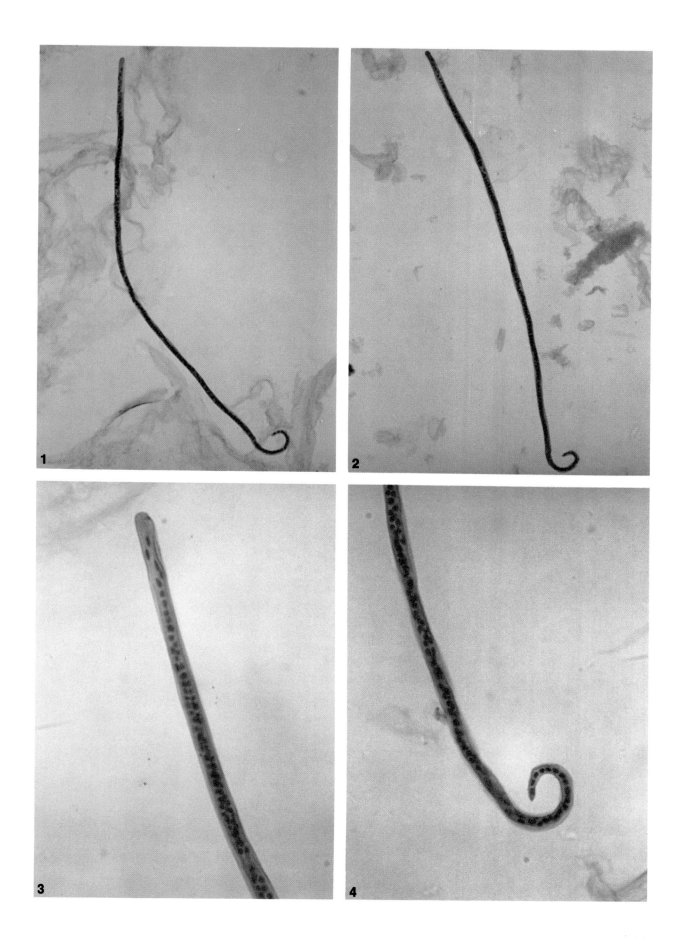

PLATE 55

Dipetalonema streptocerca,
Microfilariae in Sections of Skin

Fig 1. Portions of microfilariae *(arrow)* can be seen coiled in the superficial layer of the dermis.

Figs 2 and 3. The microfilariae may be specifically identified when tissue sections include the anterior and/or posterior extremities of the microfilariae. In Figure 2, the characteristic distribution of the nuclei in the anterior extremity immediately identifies the species. In Figure 3, the characteristic configuration of the tail and distribution of the nuclei within it also are key diagnostic features. (From the Armed Forces Institute of Pathology, negatives 337581-B and 337582-B, respectively.)

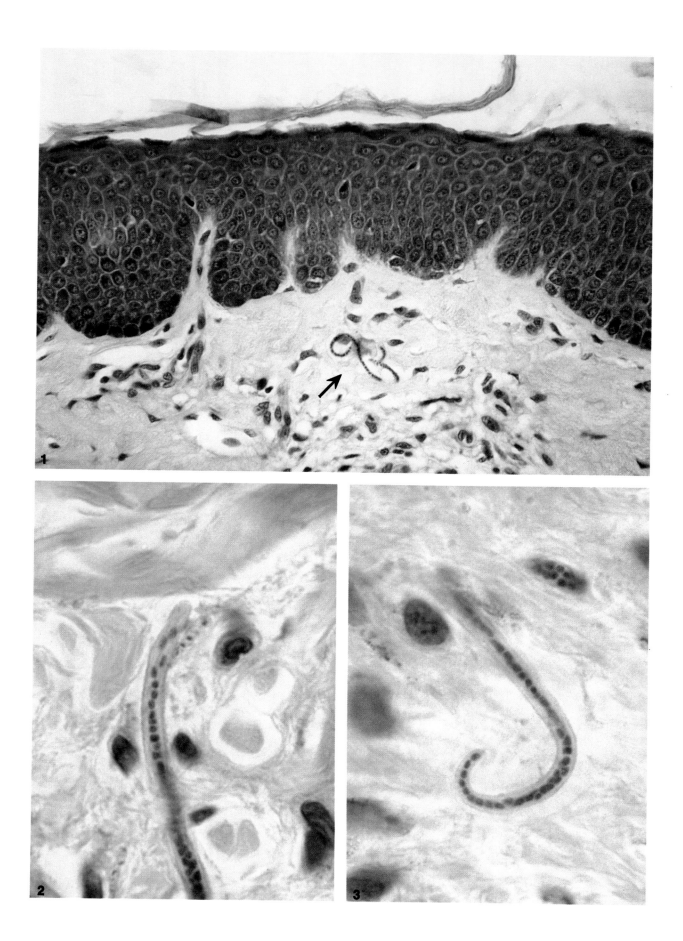

PLATE 56

Nonparasitic Elements in Blood Films

Figs 1–4. These four figures represent artifacts resembling microfilariae in size and configuration. Even though these elements take up stain in varying degrees, they lack the cellular organization characteristic of microfilariae. Various man-made fibers, fungal mycelia, and other materials may mimic microfilariae in size and shape. When in doubt, oil immersion study of these objects is required.

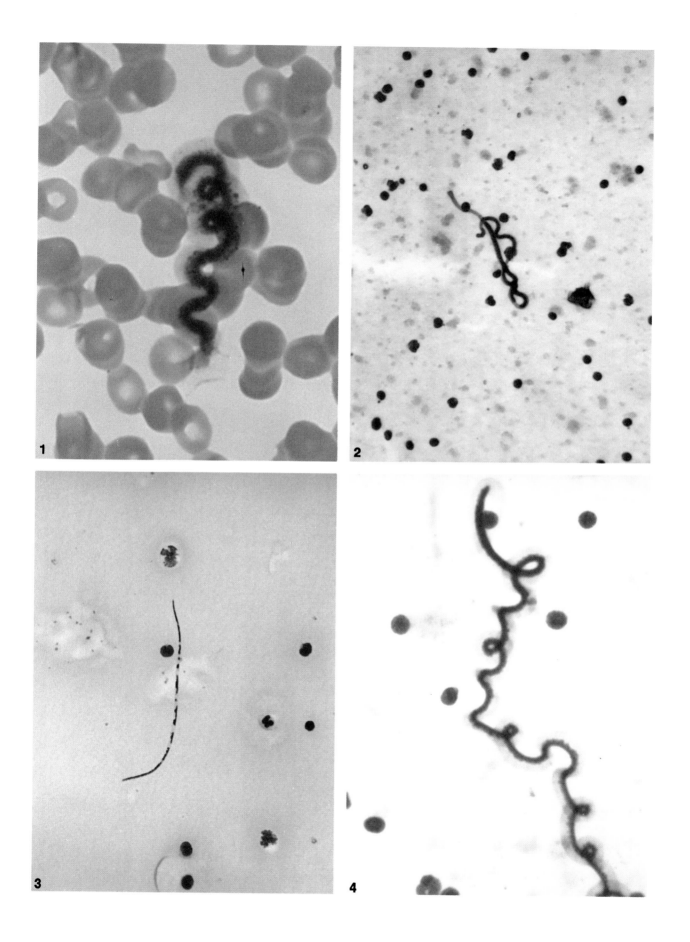

Intestinal Flukes: *Heterophyes heterophyes, Metagonimus yokogawai,* and *Echinostoma ilocanum*

CLASSIFICATION. — Trematodes.

DISEASES. — Heterophyiasis, metagonimiasis, heterophyid infection, echinostome infection.

GEOGRAPHIC DISTRIBUTION. — *H. heterophyes* occurs in the Nile River delta, Turkey, and east and southeast Asia. *M. yokogawai* is found in China, Japan, southeast Asia, and the Balkan states. *E. ilocanum* occurs in southeast Asia.

LOCATION IN HOST. — Small intestine.

MORPHOLOGY. — Adult flukes. — *H. heterophyes, M. yokogawai* and other heterophyids are all minute, pyriform flukes, measuring 1.0 – 2.5 mm long by 0.3 – 0.7 mm wide. The single ovary is situated anterior to the posteriorly positioned, paired testes. *E. ilocanum* is somewhat larger than the others, measuring 2.5 – 6.5 mm long, with a characteristic, fleshy collar provided with spines surrounding the oral sucker.

Eggs. — All heterophyid eggs, including those of *H. heterophyes* and *M. yokogawai*, are small, ovoid, operculate, and yellow-brown, measuring 25 – 30 μ by 15 – 17 μ. These eggs contain a miracidium when discharged. Although echinostome eggs closely resemble those of *Fasciola* and *Fasciolopsis* in shape, they are considerably smaller, measuring 80 – 115 μ by 58 – 70 μ. Echinostome eggs are thin-shelled, unembryonated when laid, and have an inconspicuous operculum. Frequently, they display a button-like protuberance or a roughening or slight thickening of the shell at the abopercular end.

LIFE CYCLE. — Heterophyids utilize freshwater snails and fish as first and second intermediate hosts, respectively. In fish, the metacercarial stages encyst under the scales or in the skin, and humans become infected by eating raw fish. The heterophyids lack host specificity and will attain maturity in a wide range of birds and mammals, all of which may serve as reservoirs for human infection. The prepatent period of these parasites is two to three weeks, and their normal life span is up to several months.

E. ilocanum metacercariae encyst in the tissues of large, edible snails. Humans acquire infection by eating raw snails. They have a similar prepatent period and life span.

DIAGNOSIS. — Presence of eggs in feces.

Diagnostic problems. — Because the egg-laying capacity of adult heterophyid worms is limited, fecal concentration procedures may be needed to demonstrate eggs in light infections. Accurate species identification is difficult since the eggs of many heterophyid flukes, including *H. heterophyes* and *M. yokogawai*, have similar morphologic features.

Heterophyid eggs may be confused with those of the liver flukes, *Clonorchis* and *Opisthorchis*. Usually those of the latter are somewhat larger (27 – 35 μ by 12 – 19 μ), have a seated operculum, and a knob at the end of the shell opposite the operculum. Although echinostome eggs may be mistaken for those of *Fasciola* or *Fasciolopsis*, the former's smaller size should preclude this error.

FIGURES. — Plate 57:1 – 6; also see Plate 63:1.

PLATE 57

Intestinal Fluke Eggs

Figs 1 and 2. The eggs of *H. heterophyes* are small, have a dark-staining shell, and an inconspicuous operculum. These eggs are embryonated, containing a ciliated miracidium, when passed in feces. Although they resemble *C. sinensis* eggs in size (see Fig 6), these eggs lack the shoulders and the knob at the abopercular end. The adult parasite of this species is illustrated in Plate 63, Figure 1.

Figs 3 and 4. The embryonated eggs of the small fluke, *M. yokogawai*, are about the same size as those of *Heterophyes* and *Clonorchis*. However, *M. yokogawai* eggs have a more obvious operculum than do *Heterophyes* eggs, and they lack the shoulders and knob seen in *Clonorchis* eggs. As evident in these two figures, there may be some variation in shape.

Fig 5. The egg of *E. ilocanum*, which is much larger than the other eggs illustrated here, has a thin shell, a small, inconspicuous operculum, and is unembryonated when passed in feces. This egg has an appearance similar to those of *Fasciola hepatica* and *Fasciolopsis buski* but it is much smaller.

Fig 6. This typical *C. sinensis* egg is shown for comparison of size and morphologic features with the heterophyid eggs illustrated in this Plate. A more complete description of eggs of this species is given in Plate 58.

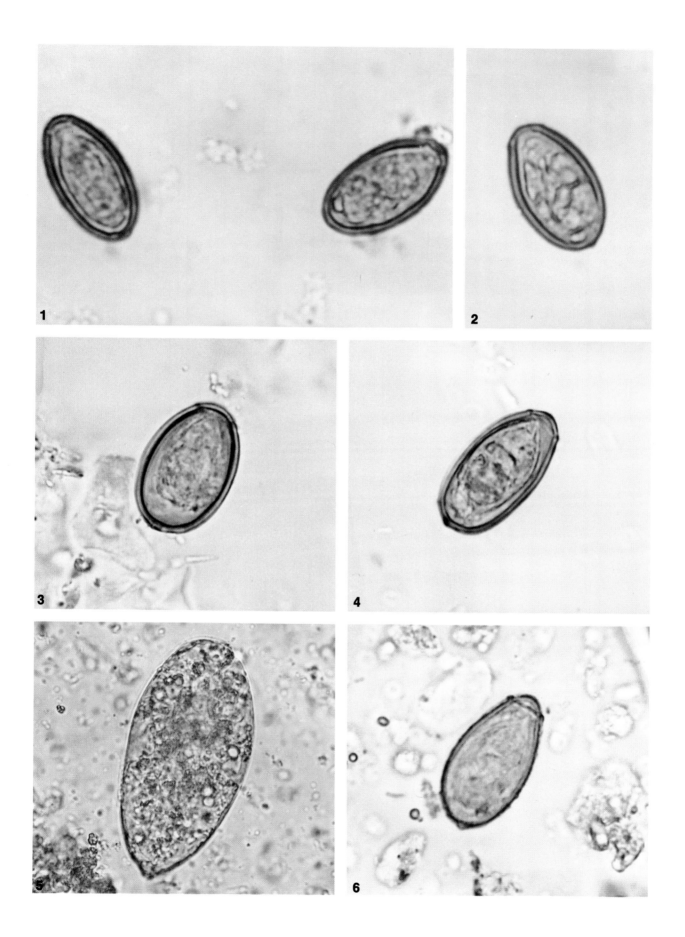

Clonorchis sinensis and Other Liver Flukes

CLASSIFICATION. — Trematodes.

DISEASE. — Clonorchiasis, Chinese liver fluke infection.

GEOGRAPHIC DISTRIBUTION. — Orient.

LOCATION IN HOST. — Bile ducts of the liver.

MORPHOLOGY. — Adult worms. — Adult worms are flattened and spatulate, measuring 10–25 mm long by 3–5 mm wide. They are hermaphroditic, with the single, rounded ovary situated anterior to the two branched testes.

Eggs. — These ovoid, yellow-brown eggs have a moderately thick shell with a seated operculum. They measure 27–35 μ by 12–19 μ, and have a small knob at the abopercular end. The eggs contain a miracidium when passed in feces.

LIFE CYCLE. — Embryonated eggs pass in feces into water where they must be ingested by an appropriate freshwater snail, the first intermediate host. Cercariae emerge from the snail and come in contact with fish, penetrating and becoming encysted under the scales or in the skin. Infection is acquired by ingestion of raw, infected fish. When metacercariae are ingested by the mammalian host, the young flukes migrate into the bile ducts of the liver via the common bile duct and attain maturity in three to four weeks. The life span of adult worms may be as long as 20–25 years. Cats and dogs serve as animal reservoirs for human infection.

DIAGNOSIS. — Detection of eggs in feces.

Diagnostic problems. — These eggs sometimes are confused with heterophyid eggs, but *Clonorchis* eggs generally are somewhat larger and have a seated operculum. Although *Clonorchis* eggs have a knob at the abopercular end, this knob is difficult to see in many instances, or it may in fact be absent.

COMMENTS. — In the United States, infections occurring in residents of Hawaii and the West Coast who have not been to the Orient probably are due to the eating of imported pickled fish in which the metacercariae have remained viable.

Other species of liver flukes belonging to the closely related genus *Opisthorchis* (eg, *O. felineus* and *O. viverrini*) also parasitize humans living in parts of southeast Asia and in northern Europe.

Another liver fluke of sheep, cattle, and other herbivorous mammals that may infect humans is *Dicrocoelium dendriticum*. Although this fluke has a world-wide distribution, most human cases occur in Asia, Europe, and Africa. The life cycle is a terrestrial one involving land snails as first intermediate hosts and ants as second intermediate hosts. The operculate eggs are quite distinctive due to their deep golden-brown color, measuring 38–45 μ by 22–30 μ. Spurious infections must be ruled out if these eggs are found in human feces.

FIGURES. — Plate 58:1, 2, 5, and 6; also see Plate 57:6 and Plate 63:2–4.

PLATE 58

Clonorchis sinensis, Opisthorchis viverrini, and Dicrocoelium dendriticum, Eggs

Figs 1, 2, and 5.   The eggs of the liver fluke, *C. sinensis*, are small and bile-stained, with a prominent operculum recessed into the shell that produces the "shoulders" characteristic of eggs of this species. Although the knob at the abopercular end (Fig 1) usually is present, it may be absent (Fig 2), or inconspicuous (Fig 5). These eggs are embryonated when passed in feces. The adult parasite is illustrated in Plate 63, Figure 4.

Figs 3 and 4.   The eggs of *O. viverrini* are difficult and frequently impossible to distinguish from those of *C. sinensis*. In *Opisthorchis*, eggs generally tend to be somewhat broader and the shoulders may be less prominent. The knob at the abopercular end, as in *Clonorchis*, may be prominent, inconspicuous, or absent. The egg in Figure 3 is slender, closely resembling *Clonorchis*, whereas the egg in Figure 4 is the more typical, broadly ovoid egg.

Fig 6.   The eggs of the liver fluke *D. dendriticum*, a relatively uncommon parasite of humans, have a thick, darkly stained shell, and an inconspicuous operculum. This egg contains a miracidium when passed in feces. The adult parasite is illustrated in Plate 63, Figure 3.

146

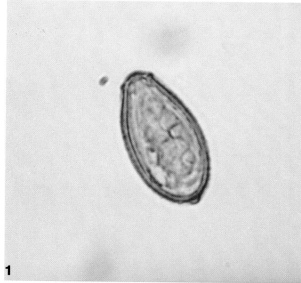

1

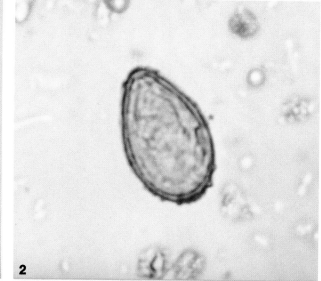

2

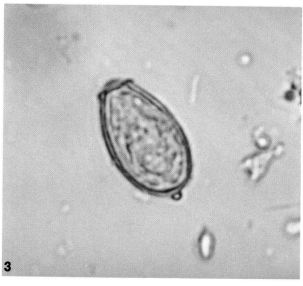

3

4

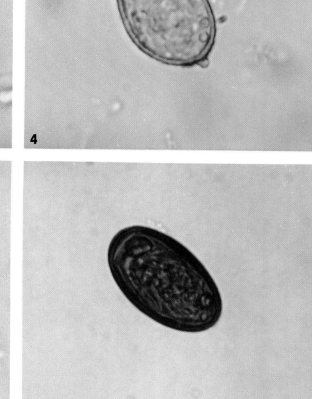

5

6

Fasciola hepatica and *Fasciolopsis buski*

CLASSIFICATION. — Trematodes.

DISEASES. — Fascioliasis, sheep liver fluke infection; fasciolopsiasis.

GEOGRAPHIC DISTRIBUTION. — *F. hepatica* has a cosmopolitan distribution, especially in sheep-raising countries. *F. buski* is found in China, Taiwan, Thailand, Indonesia, India, and other parts of Asia.

LOCATION IN HOST. — *F. hepatica* lives in the bile ducts of the liver; *F. buski* lives attached to the wall of the small intestine.

MORPHOLOGY. — Adult worms. — *F. hepatica* adults are large, fleshy worms measuring up to 30 mm long and 13 mm wide. The anterior end is distinctly cone-shaped. All internal organs — including esophagus, intestinal ceca, and reproductive organs — are extensively branched. The paired testes lie one behind the other, posterior to the ovary.

F. buski adults are also large worms measuring up to 75 mm long and 20 mm wide. The anterior end is rounded, not cone-shaped. Although the reproductive organs are extensively branched, the esophagus and intestinal ceca are not.

Eggs. — Eggs of both species are large, broadly ellipsoidal, unembryonated, and operculate. They measure $130-150 \mu$ by $63-90 \mu$ when passed in feces. The operculum is small and indistinct.

LIFE CYCLES. — The life cycles of *F. hepatica* and *F. buski* are similar. Unembryonated eggs pass in feces into water, where they must undergo embryonation and subsequent hatching of miracidia, which then seek appropriate freshwater snails as intermediate hosts. Cercariae emerge from the snails, attach to aquatic vegetation, such as watercress and water caltrop nuts, and then undergo encystation. Infection is acquired by ingestion of metacercariae on this aquatic vegetation.

Larval *F. hepatica* migrate from the intestine through the body cavity to the parenchyma of the liver; then they enter bile ducts and mature. Egg production begins in three to four months. Adult worms may live for approximately a year. Sheep and cattle serve as reservoir hosts.

F. buski matures in the duodenum in approximately three months. Pigs are an important reservoir host.

DIAGNOSIS. — Detection of eggs in feces.

Diagnostic problems. — The eggs of *F. hepatica* are indistinguishable from the eggs of *F. buski.* In areas of the Orient where these two species overlap, clinical evaluation of symptoms is an essential aid in the diagnosis of these fasciolid eggs.

FIGURES. — Plate 59:1–4; also see Plate 64:3 and Plate 64:1.

PLATE 59

Fasciola hepatica, Eggs

Figs 1–4. The large, elongate eggs of *F. hepatica* have a thin, bile-stained shell and an operculum that often is inconspicuous. They are unembryonated when passed in feces. The intestinal fluke, *Fasciolopsis buski*, produces an egg that essentially is indistinguishable from that of *F. hepatica*, and therefore has not been illustrated separately. Pressure applied to a coverslip preparation frequently may result in "popping" the operculum (Fig 4). The adult parasites of *F. buski* and *F. hepatica* are illustrated in Plate 64, Figures 1 and 3, respectively.

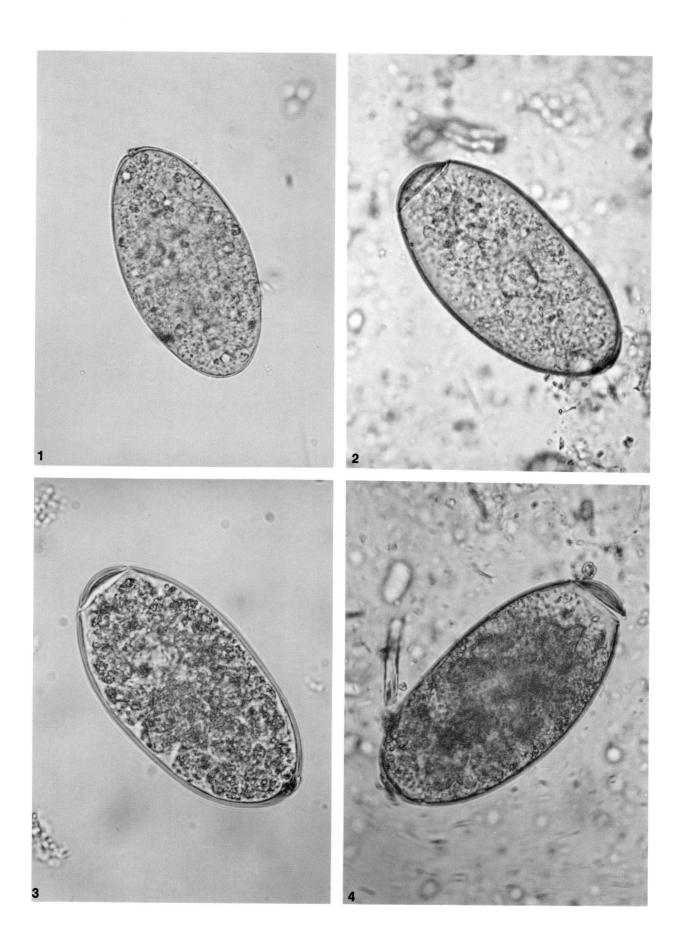

Paragonimus westermani

CLASSIFICATION. — Trematode.

DISEASE. — Paragonimiasis, Oriental lung fluke infection.

GEOGRAPHIC DISTRIBUTION. — Asia.

LOCATION IN HOST. — Encapsulated in parenchyma of lung.

MORPHOLOGY. — Adult worms. — These thick, robust, ovoid flukes, measuring 7.5 – 12.0 mm by 4 – 6 mm, are hermaphroditic, with a lobed ovary located anterior to two branched testes. The testes lie side by side in the posterior part of the body.

Eggs. — The moderately large, broadly ovoid, thick-shelled, golden-brown eggs are unembryonated when passed in sputum or in feces. The eggs may measure 80 – 120 μ by 45 – 70 μ. At the abopercular end of the egg, the shell is somewhat thickened.

LIFE CYCLE. — Unembryonated eggs passed in feces enter the water and undergo development to the miracidial stage. The egg hatches and the miracidium must find an appropriate freshwater snail intermediate host. Following a developmental period of two to three months, cercariae emerge from the snail and come in contact with various crustacea, such as crabs and crayfish, penetrating into the viscera and musculature, and encysting to the metacercarial stage. Infection is acquired by ingestion of raw, infected crustaceans.

In the mammalian host, the metacercariae migrate from the intestine through the body cavity and the diaphragm where they then enter the parenchyma of the lung. Adult worms become encapsulated in lung tissue, usually two worms per each capsule. The prepatent period is five to six weeks, and adult worms may live for up to 20 years.

DIAGNOSIS. — Detection of eggs in feces or in sputum.

Diagnostic problems. — The eggs of this species frequently are confused with the eggs of the broad fish tapeworm, *Diphyllobothrium latum*. However, the eggs of the latter are smaller (58 – 75 μ long by 40 – 50 μ) and they have an abopercular knob on the shell. In *Paragonimus*, the abopercular end of the thickened shell has no knob.

COMMENTS. — It now is well recognized that a number of *Paragonimus* species may parasitize humans, and almost all of these occur in canine or feline reservoir hosts. *P. westermani* matures in dogs and cats. The other species of *Paragonimus* have been described from various parts of southeast Asia, Africa, and Latin America. The eggs of these other species, though varying in size, have generally similar morphology.

FIGURES. — Plate 60:1 – 6; also see Plate 64:2.

PLATE 60

Paragonimus westermani, Eggs

Figs 1–4 and 6. The eggs of the lung fluke, *P. westermani*, are large, have a moderately thick, dark shell, a prominent operculum at the broad end, and a thickened abopercular end. They are unembryonated when discharged from the body in feces or in sputum. These eggs illustrate the wide variation in shape that is characteristic of the species. The adult parasite is illustrated in Plate 64, Figure 2.

Fig 5. The eggs of *P. westermani* sometimes are confused with those of the tapeworm *D. latum*. However, as shown in this illustration of both species in the same microscopic field, the *Paragonimus* egg is larger, has a thicker and more deeply stained shell, and the thickened abopercular end lacks the knob so often seen in *D. latum*.

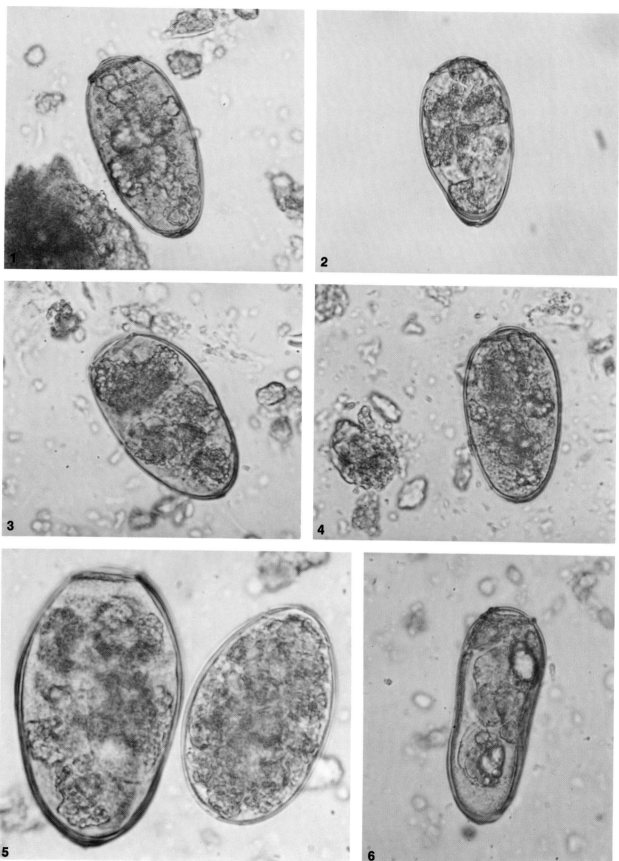

Schistosoma mansoni

CLASSIFICATION. — Trematode.

DISEASE. — Schistosomiasis mansoni, intestinal bilharziasis.

GEOGRAPHIC DISTRIBUTION. — Africa, Arabian peninsula, West Indies, Puerto Rico, Brazil, Venezuela, and Surinam.

LOCATION IN HOST. — Venous plexuses of the colon and lower ileum; portal system of the liver.

MORPHOLOGY. — Adult worms. — Sexes are separate. Male worms measure 6.4–12.0 mm long, and females are 7.2–17.0 mm long. Male worms are robust, and the posterior portion of the body is folded to form a gynecophoral canal within which the more slender female worm lies in permanent copula. Males have a grossly tuberculate integument, and the testes — from six to nine in number — are located in the anterior half of the body. Females typically have only a few eggs in their uterus.

Eggs. — Eggs, measuring 114–175 μ by 45–70 μ, are large and nonoperculate, with a transparent shell and prominent lateral spine. They contain a miracidium when passed in feces.

LIFE CYCLE. — Embryonated eggs pass in feces into water where the miracidia hatch and swim about until the appropriate snail intermediate host (*Biomphalaria* spp) is invaded. Following development in the snail, cercariae emerge, swim freely in water, and then directly invade the skin of the human host. Animal reservoirs of infection are not important for this species. When the tail of the cercaria is lost upon invasion of the skin, the larval stage — referred to as a *schistosomulum* — migrates through the lungs and liver. Eventually, the parasites develop to maturity in the superior and inferior mesenteric veins. The prepatent period is six to eight weeks. Adult worms may live for up to 25 years.

DIAGNOSIS. — Presence of eggs in feces and in rectal biopsy specimens.

Diagnostic problems. — The identification of these eggs usually poses no difficulty because of the large lateral spine. However, since very few eggs may pass in feces, concentration methods or rectal biopsy may have to be performed.

COMMENTS. — Although schistosomiasis is not endemic in the United States, *S. mansoni* infections frequently may be seen in urban centers such as New York, Chicago, and other areas with high immigrant populations from Puerto Rico, the West Indies, and other parts of Latin America.

FIGURES. — Plate 61:1–4; also see Plate 64:4.

PLATE 61

Schistosoma mansoni, Eggs

Figs 1–4. The egg of this schistosome is large and elongate, has a thin shell and a prominent lateral spine, and contains a miracidium when passed in feces. In coverslip preparations, the lateral spine occasionally will be hidden from view (Fig 2). Light tapping on the coverslip usually will reorient the egg and expose the spine. In iodine-stained preparations (Fig 3), the miracidium usually stains darkly. In fresh saline preparations, the miracidium frequently is seen moving in the egg.

Although not shown here, eggs also may be detected in rectal biopsy specimens. They have the same appearance in both cases. By observing movement of the miracidium or of the flame cells within the organism, one can determine whether eggs are alive or dead. The adult parasites are illustrated in Plate 64, Figure 4.

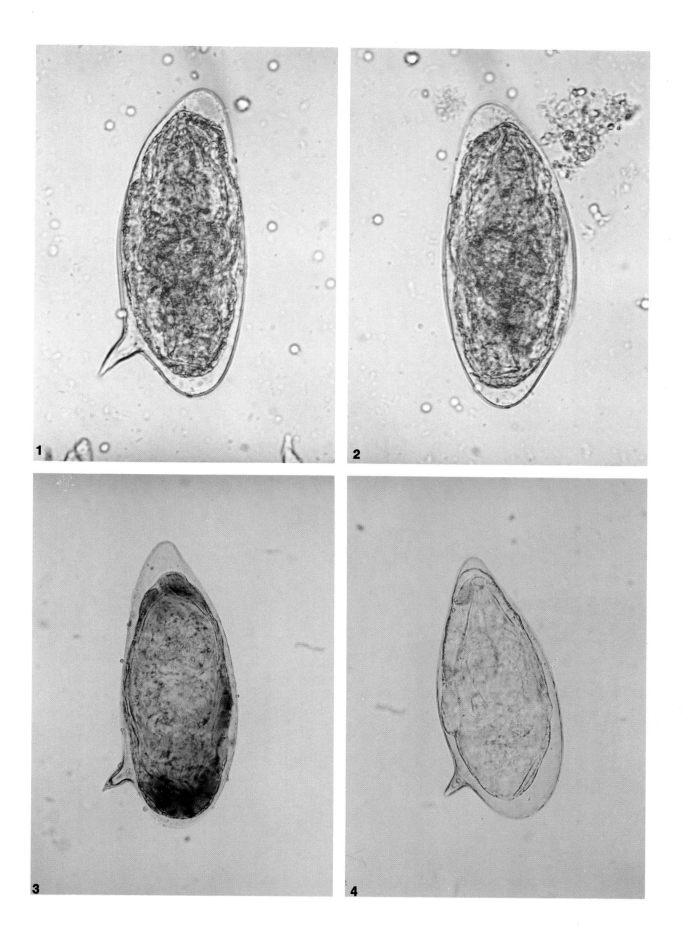

1

2

3

4

Schistosoma japonicum, S. *mekongi,* and S. *haematobium*

CLASSIFICATION.—Trematodes.

DISEASES.—Schistosomiasis japonica or Oriental blood fluke infection; Mekong River schistosomiasis, schistosomiasis haematobia, or urinary bilharziasis.

GEOGRAPHIC DISTRIBUTION.—S. *japonicum* occurs in China, the Philippines and other countries of southeast Asia. S. *mekongi* occurs along the Mekong River in southeast Asia. S. *haematobium* is found in Africa, the Arabian peninsula, Iran, Syria, Lebanon, and Malagasy.

LOCATION IN HOST.—S. *japonicum* and S. *mekongi* live in the venous plexuses of the small intestine, and S. *haematobium* is found in the venous plexuses of the bladder and rectum.

MORPHOLOGY.—Adult worms.—As in S. *mansoni*, the sexes are separate and their sizes are comparable. S. *japonicum* and S. *mekongi* will contain more eggs in utero than either S. *mansoni* or S. *haematobium*.

Eggs.—S. *japonicum* eggs are large, round, and nonoperculate, measuring 70–100 μ by 55–65 μ. They have a transparent shell with a small, inconspicuous spine. S. *mekongi* eggs are similar to those of S. *japonicum* except the former is smaller—50–65 μ by 30–55 μ. Eggs of S. *haematobium* are large, nonoperculate, and have a transparent shell with a prominent terminal spine; they measure 112–170 μ by 40–70 μ.

LIFE CYCLE.—The life cycles of these three species are similar to that of S. *mansoni* except that the intermediate snail hosts differ, as does the final habitat of the adult worms. Snails of the genus *Oncomelania*, are the intermediate hosts for S. *japonicum*. *Tricula* snails are intermediate hosts for S. *mekongi*, and *Bulinus* snails serve as intermediate hosts for S. *haematobium*. The prepatent periods for S. *japonicum* and S. *mekongi* are from four to six weeks, whereas the prepatent period is 8–12 weeks for S. *haematobium*.

DIAGNOSIS.—S. *japonicum* and S. *mekongi* are diagnosed by the demonstration of eggs in feces or rectal biopsy specimens. Eggs of S. *haematobium* usually are found in urine.

Diagnostic problems.—Frequently, considerable fecal debris adheres to the shells of S. *japonicum* and S. *mekongi* eggs; this may result in the eggs being overlooked in fecal preparations. In addition, the spine often is difficult to see. In S. *haematobium*, it usually is necessary to concentrate urine to detect eggs.

FIGURES.—Plate 62:1–7.

PLATE 62

Schistosoma japonicum, S. *mekongi,* and S. *haematobium,* Eggs

Figs 1 and 2. The S. *japonicum* egg is smaller than those of S. *mansoni* and S. *haematobium*. It is ovoid and has a thin, clear shell and a small spine or hook on its lateral margin that may be inconspicuous and difficult to see (Fig 1). This egg, like the other schistosome eggs, contains a miracidium when passed in feces. The eggs of this species frequently have fecal debris adhering to their surface, making their identification more difficult.

Figs 3 and 4. The eggs of S. *mekongi*, a recently described schistosome, closely resemble those of S. *japonicum*. Although S. *mekongi* eggs are smaller, they are similar to the latter in most other respects. The lateral spine tends to be more conspicuous than in S. *japonicum*.

Figs 5–7. The eggs of S. *haematobium* occur in the urine. They are large, approximately the same size as those of S. *mansoni*, and have a terminal spine. As shown in these figures, there may be variations in the shape of the egg and in the character of the spine.

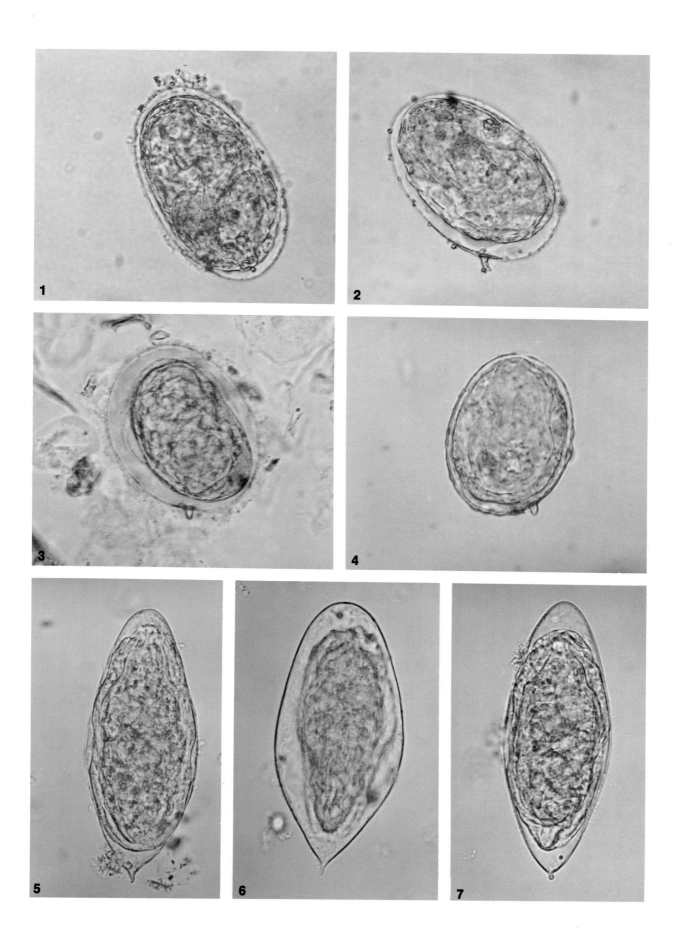

PLATE 63

Adult Trematodes

Fig 1. *Heterophyes heterophyes,* adult worm, carmine stain. This minute fluke of the human small intestine measures 1–2 mm long. The ventral sucker is larger than the oral sucker. The paired testes lie side by side at the extreme posterior end and the small ovary is situated anterior to them. The dark brown-staining yolk glands are few in number in the posterior portion of the worm along the lateral margin. Minute spines cover the surface of the worm.

Fig 2. *Opisthorchis viverrini,* adult worm, carmine stain. This fluke lives in the bile passages of the liver, and has a usual length of 10–25 mm. Its structure closely resembles that of *Clonorchis sinensis* (Fig 4), with the two testes lying one behind the other in the posterior portion of the body. The ovary is anterior to the testes and the brown-staining uterus forms many coils between the ovary and the ventral sucker. Vitellaria are along the lateral margins in the midportion of the worm. The testes are not as deeply lobed as in *Clonorchis.*

Fig 3. *Dicrocoelium dendriticum,* adult worm, carmine stain. This liver fluke occurs in the bile ducts of sheep, goats, and occasionally humans. It measures about 5–15 mm long and has a smooth cuticle. The testes lie side by side in the anterior part of the body just below the oral sucker. The ovary is posterior to the testes, and the uterine coils fill in the posterior portion of the body. Yolk glands are along the lateral margins in the midportion of the body.

Fig 4. *Clonorchis sinensis,* adult worm, carmine stain. These adults live in the bile passages of humans and various animals, and usually measure 10–25 mm long. The two highly branched testes lie one behind the other in the posterior portion of the worm. The ovary is anterior to the testes, and the coiled uterus fills in the midportion of the body between the ovary and the ventral sucker. Vitelline glands are along the lateral margins in the midportion of the fluke. There are no spines on the surface.

NOTE: These adult flukes are not illustrated to scale because of the extreme size differences among species.

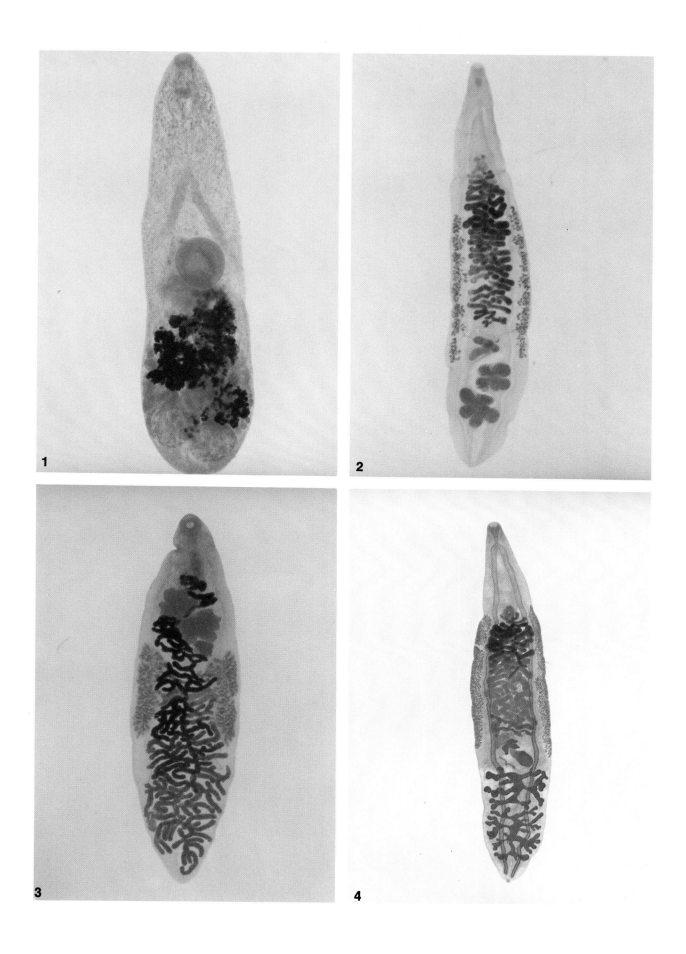

PLATE 64

Adult Trematodes

Fig 1. *Fasciolopsis buski,* adult worm, carmine stain. This very large fluke, which lives in the intestine of pigs and humans and usually measures 20–75 mm long by 0.5–3.0 mm thick, has poorly developed suckers. The extensively branched, pink-staining testes occupy most of the posterior half of the worm. The branched ovary is located anterior to the testes and the short, coiled uterus lies between the ovary and the ventral sucker. The brown-staining yolk glands lie along the lateral margins of the worm. The surface of the body is spinous.

Fig 2. *Paragonimus westermani,* adult fluke, carmine stain. This small, robust trematode—known as the Oriental lung fluke—lives encapsulated in the lungs of humans and other animals. It usually measures 7.5–12.0 mm long by 3–5 mm thick. The testes are relatively large and branched, lying side by side in the posterior part of the body. The ovary is somewhat lobed and is anterior to the testes. The uterus is dark-staining in the center of the worm, and the brown yolk glands occupy the lateral field, filling in toward the center of the worm.

Fig 3. *Fasciola hepatica,* adult fluke, carmine stain. This large liver fluke of humans and herbivorous animals lives in the bile ducts, and usually measures up to 33 mm long. It has a characteristic, cone-shaped anterior end, with extensive branching of the intestine and reproductive organs. The dark-staining uterus is in the anterior third of the worm, and the yolk glands are the brown-staining structures occupying the lateral fields. The surface of the parasite is spinous.

Fig 4. *Schistosoma mansoni,* male and female adults, carmine stain. The schistosomes are the only human trematodes with separate sexes. The long slender female lies within the gynecophoral canal of the more robust male. Several small testes in the anterior portion of the male worm may be seen. These flukes live in blood vessels of the human host, as do all the other schistosomes.

NOTE: These adult flukes are not illustrated to scale because of the extreme size differences among species.

Diphyllobothrium latum

CLASSIFICATION. — Cestode.

DISEASE. — Diphyllobothriasis, fish tapeworm infection.

GEOGRAPHIC DISTRIBUTION. — Most common in temperate regions where cold, clear lakes are abundant; especially common in northern Europe, the Baltic countries, North America, and Japan.

LOCATION IN HOST. — Small intestine.

MORPHOLOGY. — Adult worms. — The adult tapeworm may be 4–10 meters long. The scolex is small (3 × 1 mm) and spatulate, with two shallow grooves (bothria). Gravid proglottids are wider than they are long (3 × 11 mm) and have a rosette-shaped central uterus, with the genital pore situated on the midventral surface.

Eggs. — The ovoid, operculate, moderately thick-shelled, yellowish-brown eggs measure 58–75 μ by 40–50 μ. There usually is a small knob at the end opposite the operculum, however, it may be indistinct in many instances. The egg is undeveloped when passed in feces.

LIFE CYCLE. — The unembryonated eggs pass in feces into water where they incubate for approximately two weeks, and then contain a ciliated six-hooked embryo called a *coracidium*. The coracidium hatches in water, and it then may be ingested by a copepod, the first intermediate host. Within the body cavity of the copepod, there is development to a solid-bodied larva called a *procercoid*. When this stage is ingested by a fish, the second intermediate host, the larva moves into the flesh, growing into a larval stage known as a *plerocercoid* (or *sparganum*). Ingestion of the sparganum by humans or dogs results in development to the adult tapeworm within a period of three to five weeks. This tapeworm may live 25 years or longer.

DIAGNOSIS. — By demonstration of eggs in feces; sometimes proglottids are found in feces.

Diagnostic problems. — These eggs are sometimes confused with those of *Paragonimus westermani*, the human lung fluke. The eggs of *D. latum* are smaller than those of *P. westermani*. The latter are always more than 80 μ long. Also, the abopercular end of the shell of *P. westermani* is thickened but does not have the knob usually seen on *D. latum* eggs.

FIGURES. — Plate 65:1–4; Plate 66:1–5; also see Plate 60:5.

PLATE 65

Diphyllobothrium latum, Adult Worms

Figs 1 and 2. The scolex of the broad fish tapeworm is characterized by its two lateral grooves, known as bothria, rather than the four suckers observed in the other human tapeworms. The bothria are well illustrated in the preserved but unstained scolex shown in Figure 1. In a stained preparation, (Fig 2) the grooves *(arrow)* are less well defined.

Figs 3 and 4. Mature proglottids of this species are broader than long, with a centrally positioned genital pore. The coiled uterus has the appearance of a rosette. Figure 4 illustrates the structure of the uterus at a higher magnification. In this infection, eggs usually are passed in feces. However, the tapeworm may shed old proglottids, and it is not uncommon to find a chain of proglottids in fecal specimens.

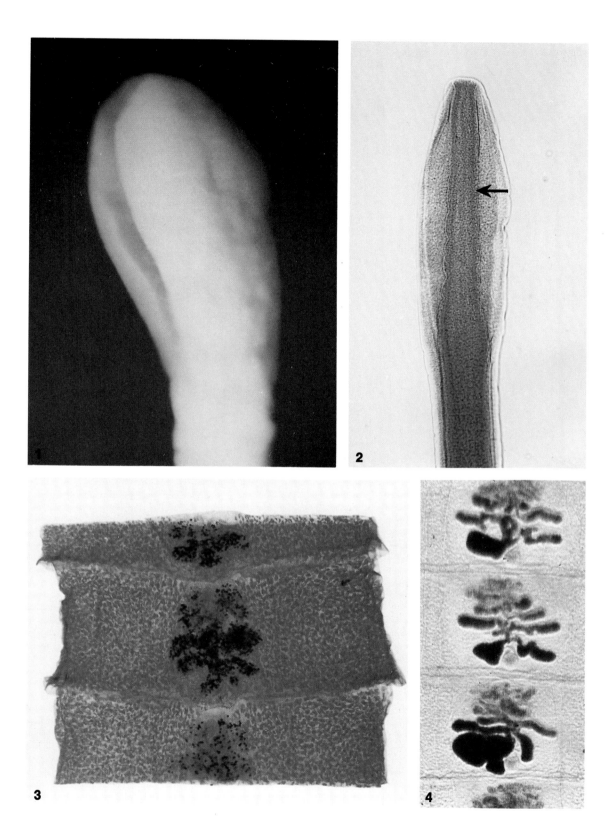

PLATE 66

Diphyllobothrium latum, **Eggs**

Figs 1–3. The eggs of this tapeworm are discharged from the proglottids, and pass out in feces in an unembryonated condition. The egg is ovoid and moderately large, with a thin, bile-stained shell, an inconspicuous operculum, and a knob at the abopercular end. The knob may not always be clearly evident; in many instances it appears to be absent. The presence of an operculum often leads to misidentification of this egg as a trematode egg. Although this egg is confused most often with *Paragonimus westermani,* the latter is always larger and has a thickened abopercular end. *P. westermani* lacks the knob seen in *D. latum.* A *Paragonimus* egg and a *D. latum* egg in the same field are illustrated in Plate 60, Figure 5.

Figs 4 and 5. At high magnification, the operculum (Fig 4) and the abopercular knob (Fig 5) are clearly seen.

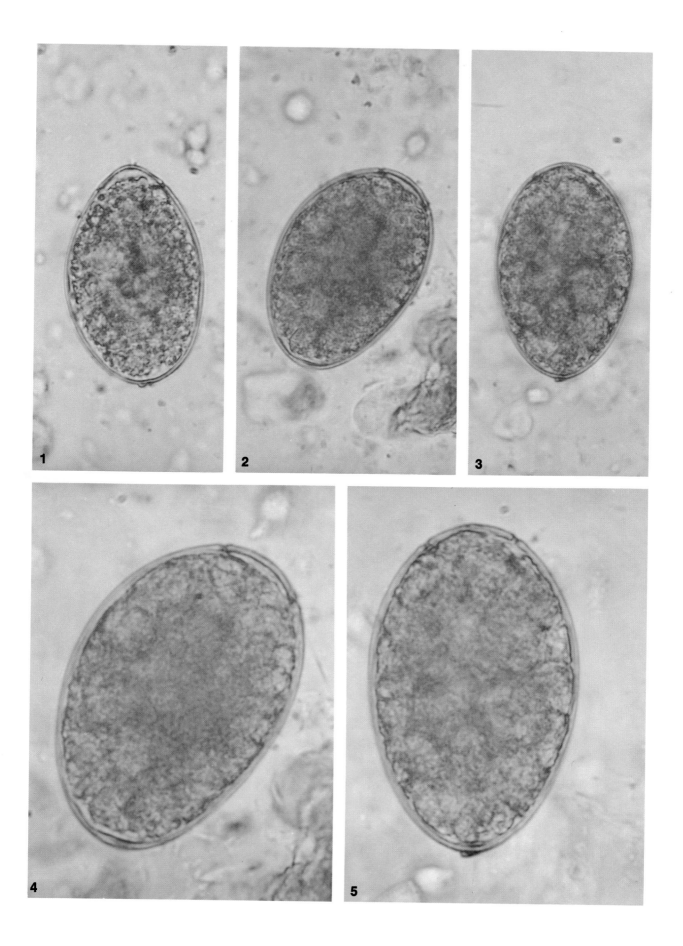

Taenia saginata

CLASSIFICATION. — Cestode.

DISEASE. — Taeniasis, beef tapeworm infection.

GEOGRAPHIC DISTRIBUTION. — Cosmopolitan.

LOCATION IN HOST. — Small intestine.

MORPHOLOGY. — Adult worms. — The adult tapeworm may attain a length of 4 – 8 meters. The scolex is small (1 – 2 mm in diameter) and has four suckers. There is no rostellum, nor are there hooklets. Gravid proglottids are longer than they are wide (18 – 20 mm by 5 – 7 mm), and have 15 – 20 lateral branches on each side of the central uterine stem. A genital pore is located at the lateral margin.

Eggs. — The spherical, yellow-brown eggs measure 31 – 43 μ. The shell is thick and radially striated, giving it a prismatic appearance. Occasionally, a thin outer primary membrane can be seen around some eggs, especially if a proglottid is ruptured. The egg contains a six-hooked embryo called an oncosphere.

Larval stage. — The larval stage occurring in the tissues of the intermediate host is referred to as a *cysticercus*. It is a small (diameter, 0.5 – 2.0 mm), rounded, opalescent structure that contains an unarmed scolex with four suckers inverted into a fluid-filled bladder. Cysticerci of this species — referred to as *Cysticercus bovis* — are found in herbivorous animals but not in humans.

LIFE CYCLE. — Eggs liberated from gravid proglottids pass in feces into soil or water, eventually reaching pasture land where they are ingested by cattle or other herbivores. Development in the muscles to infective cysticerci takes two to three months. Following ingestion by humans, the cysticerci require three to five months to mature to adult tapeworms in the intestine. The tapeworm may live 25 years or longer. Eggs of *T. saginata* are not infective to humans.

DIAGNOSIS. — By demonstration of eggs in feces, or by finding gravid proglottids that have spontaneously emerged from the anal opening or have been passed in feces. Eggs sometimes can be demonstrated in cellulose tape preparations taken from the perianal skin.

Diagnostic problems. — The eggs of human and animal taeniid species (including *Taenia* and *Echinococcus* spp) are all identical and indistinguishable from each other. In humans, it usually is only necessary to determine whether infection is due to *T. saginata* or *T. solium*. This diagnosis is usually done by examination of gravid proglottids that have been injected with India ink or stained by permanent stains to reveal the characteristic number of lateral uterine branches. Some pollen grains that may be found in feces closely resemble *Taenia* eggs in size, color, and appearance. Differentiation is made by visualizing the six-hooked embryo present in *Taenia* eggs.

COMMENTS. — All *Taenia* eggs and proglottids *must be handled with extreme care* since the eggs of *T. solium*, indistinguishable from those of *T. saginata*, are infective to humans and may cause human cysticercosis if they are ingested.

FIGURES. — Plate 67:1 – 2; Plate 68:2 – 4.

PLATE 67

Taenia saginata and *T. solium,* Adult Worms

Figs 1 and 2.   *T. saginata*. The scolex (Fig 1) of this species has four large suckers but no rostellar hooks. The gravid proglottid of this species (Fig 2) shows the characteristic branching of the uterus and the position of the genital pore at the lateral margin. In *T. saginata*, there are 15 or more lateral branches on each side of the central uterine stem.

Figs 3 and 4.   *T. solium*. The scolex of *T. solium* (Fig 3) has, in addition to the four suckers, a rostellum bearing two rows of large hooks. The gravid proglottids (Fig 4), in contrast to those of *T. saginata*, have less than 13 lateral branches on each side of the central uterine stem; this feature serves to distinguish the two species.

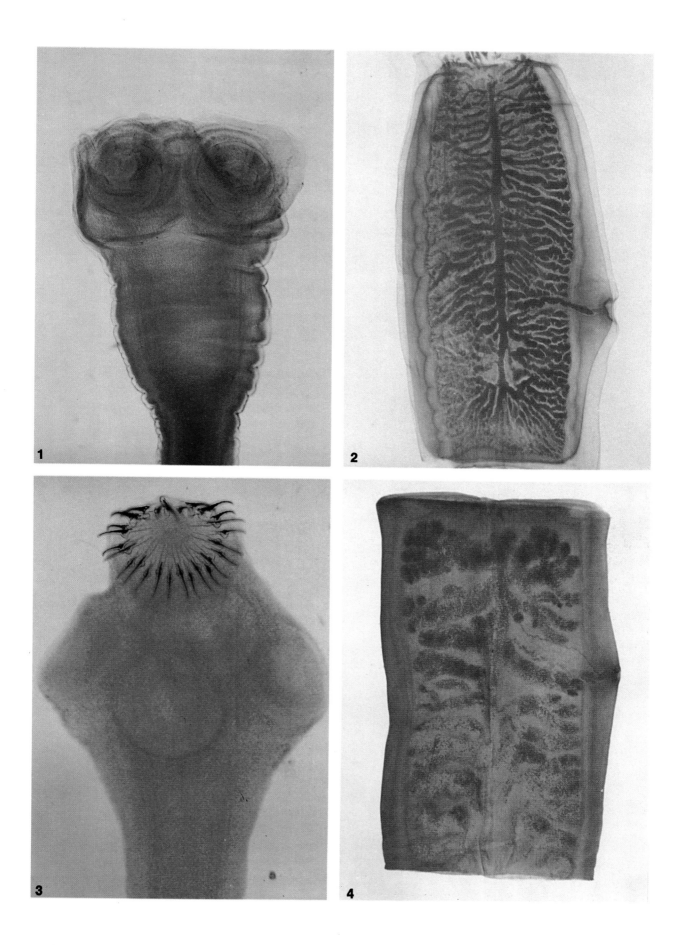

Taenia solium

CLASSIFICATION. — Cestode.

DISEASE. — Taeniasis, pork tapeworm infection.

GEOGRAPHIC DISTRIBUTION. — Cosmopolitan, but especially middle European countries, Latin America, India, and China.

LOCATION IN HOST. — Small intestine.

MORPHOLOGY. — Adult worms. — The adult tapeworms may attain lengths of 3–5 meters. The scolex is small (approximately 1 mm diameter), has four suckers, and there is a rostellum with two rows of hooklets. Gravid proglottids are longer than they are wide (11 × 5 mm), and have 7–13 lateral branches on each side of the central uterine stem. The genital pore is located at the lateral margin.

Eggs. — As described for *T. saginata* (see text preceding Plate 67).

Larval stage. — The larval stage occurring in the tissues of the intermediate host is referred to as a *cysticercus;* the larva of *T. solium* is called *Cysticer-*

cus cellulosae. It is a small (diameter, 0.5–2.0 mm), rounded, opalescent structure that contains an armed scolex inverted into a fluid-filled bladder. Cysticerci of this species are found in pigs (usual and normal intermediate host), humans, and other animals.

LIFE CYCLE. — The life cycle is the same as described for *T. saginata.* The eggs of *T. solium*, in contrast to those of *T. saginata*, are infective to humans, and, when ingested, they may result in a serious, and sometimes fatal, disease called cysticercosis.

DIAGNOSIS. — Same as described for *T. saginata.*

COMMENTS. — All *Taenia* eggs and proglottids *must be handled with extreme care* since the eggs of *T. solium*, indistinguishable from those of *T. saginata*, are infective to humans and may cause human cysticercosis if they are ingested.

Plate 68:1, 2, and 4; also see Plate 67:3–4.

PLATE 68

Taenia solium and *T. saginata,* Eggs and Proglottids

Figs 1 and 3. For rapid identification of *Taenia* proglottids passed in feces, India ink may be injected into the uterus by syringe and needle through the minute opening of the lateral genital pore. The black-staining uterus then allows for the enumeration of the primary lateral branches coming off the central uterine stem. Those parasites having 13 or fewer branches are *T. solium* (Fig 1), while those with 15 or more are *T. saginata* (Fig 3). In handling unidentified *Taenia* proglottids, caution should be used since the eggs of *T. solium* are infective to man and can cause cysticercosis.

Figs 2 and 4. Eggs of *Taenia* species have a thick, bile-stained, radially striated shell enclosing a six-hooked embryo (oncosphere). The eggs of *T. solium* and *T. saginata* are indistinguishable from each other, from eggs of *Echinococcus granulosus,* and other animal taeniid tapeworms as well.

The shell may stain so darkly with iodine that the egg resembles a pollen grain. For absolute diagnosis, the six hooks of the embryo must be seen. Occasionally, some eggs in feces or those ruptured from proglottids may retain their delicate, primary membrane (Fig 2), but more often it is absent (Fig 4). *Taenia* eggs may be more oval-shaped than round. In preserved eggs, the embryo may become quite granular and the hooks more difficult to see. In fresh specimens, the hooks are almost always evident by careful focusing on the enclosed embryo.

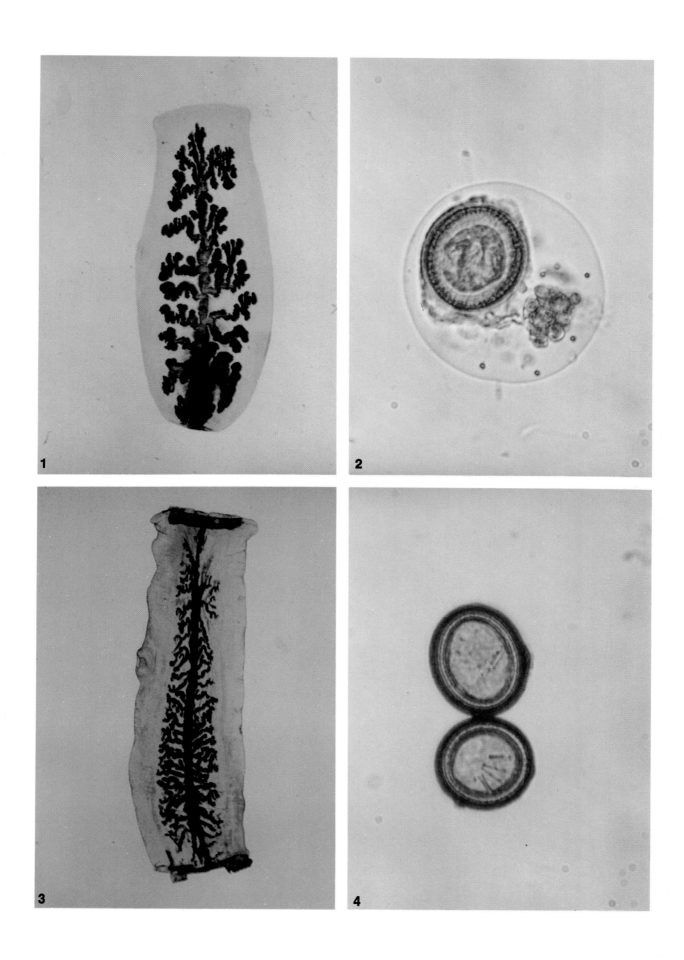

Hymenolepis nana

CLASSIFICATION. — Cestode.

DISEASE. — Hymenolepiasis, dwarf tapeworm infection.

GEOGRAPHIC DISTRIBUTION. — Cosmopolitan.

LOCATION IN HOST. — Small intestine.

MORPHOLOGY. — Adult worms. — Adult tapeworms are very small, measuring 2.5–4.0 cm long. The scolex is tiny and knob-like with four suckers and a rostellum and hooklets. Proglottids are wider than they are long.

Eggs. — The spherical to subspherical eggs have a thin, hyaline shell, and measure 30–47 μ in diameter. The six-hooked oncosphere is surrounded by a membrane that has two polar thickenings, from which arise four to eight filaments extending into the space between the embryo and the outer shell.

LIFE CYCLE. — Eggs pass in feces into the external environment where, in the usual life cycle involving rodents, the eggs are ingested by various arthropods, generally beetles, which serve as intermediate hosts. In the beetle, the oncosphere develops into the infective, larval stage called a *cysticercoid*. Human infection usually is direct, by ingestion of eggs. Human infection by ingestion of infected beetles also is possible. When infection is acquired by ingesting eggs the cysticercoid stage develops within the wall of the small intestine before emerging into the intestinal lumen to reach maturity as an adult tapeworm in two to three weeks.

Internal autoinfection also is possible, wherein eggs being passed by the adult tapeworm hatch within the intestinal tract and go through the cysticercoid stage, maturing in the intestine as adult tapeworms. Although mice are the usual hosts for this parasite, human infections are common in many areas of the world.

DIAGNOSIS. — Demonstration of characteristic eggs in feces.

Diagnostic problems. — *H. nana* eggs sometimes are confused with the eggs of *H. diminuta*, the rat tapeworm, which also may infect humans. The eggs of *H. diminuta* are much larger (70–85 μ by 60–80 μ), and do not have polar filaments arising from the membrane around the oncosphere.

COMMENTS. — This can be a very common infection in children in institutional settings.

FIGURES. — Plate 69:1–4; Plate 70:4.

PLATE 69

Hymenolepis nana, Eggs

Figs 1–3. The clear, thin-shelled eggs of *H. nana* are spherical to ovoid in shape, and contain a six-hooked embryo (oncosphere) when passed in feces. The most prominent characteristic of these eggs is the filaments that arise from polar thickenings of the inner envelope surrounding the oncosphere. With prolonged storage in formalin these filaments are difficult to see. At higher magnification (Fig 3), the minute structure of the egg can be more clearly seen.

Fig 4. The eggs of *H. nana* should not be confused with those of a related species, *H. diminuta*. When the two are shown in the same microscopic field, it is readily apparent that the egg of *H. nana* is smaller and has the polar filaments that are absent in *H. diminuta*. Additional features of *H. diminuta* are illustrated in Plate 70.

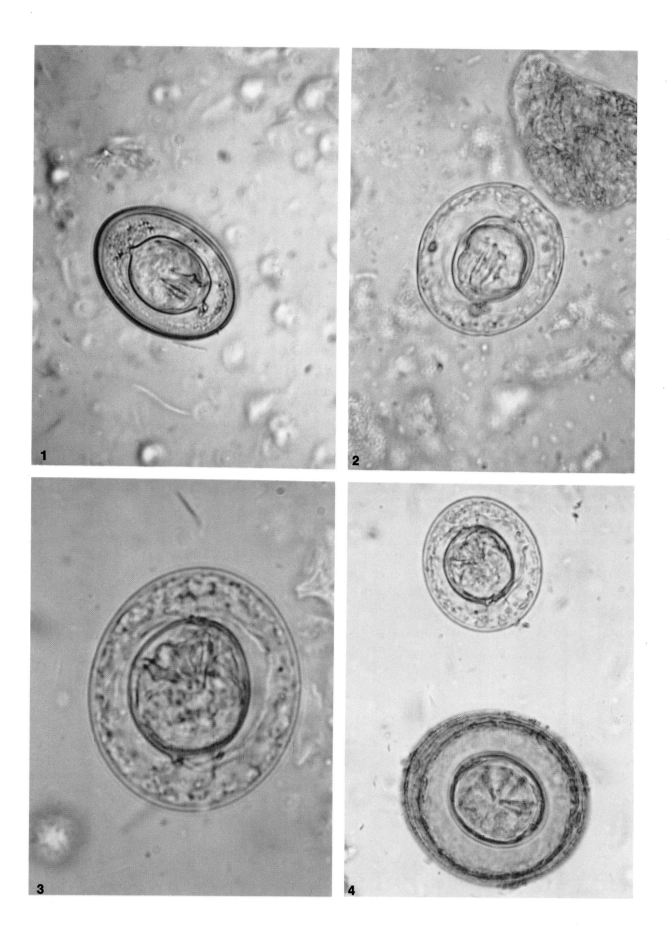

Hymenolepis diminuta

CLASSIFICATION. — Cestode.

DISEASE. — Rat tapeworm infection.

GEOGRAPHIC DISTRIBUTION. — Cosmopolitan.

LOCATION IN HOST. — Small intestine.

MORPHOLOGY. — Adult worms. — Adult tapeworms are from 20–60 cm long. The scolex is knoblike, has four suckers, and has a rostellum but no hooklets. Proglottids are wider than they are long.

Eggs. — The large, spherical, yellowish-brown, thick-shelled eggs measure 70–85 μ by 60–80 μ. The six-hooked oncosphere is surrounded by a membrane that is considerably separated from the outer shell. There are no polar filaments extending into the space between the oncosphere and outer shell.

LIFE CYCLE. — The life cycle of this species is similar to that of *H. nana*. Eggs passed in feces must be ingested by an arthropod intermediate host, usually a beetle. In the arthropod, the oncosphere de-velops into a larval stage called a cysticercoid which is then infective to rats and to humans. Human infection usually is acquired by accidental ingestion of infected beetles present in various grain products. Direct infection by eggs, without use of an arthropod intermediate host as occurs in *H. nana*, is not possible. Maturation to the adult tapeworm in the small intestine takes approximately three weeks. Rats are the usual host for this parasite but human infections are not infrequent. Internal autoinfection does not occur with this species.

DIAGNOSIS. — Demonstration of characteristic eggs in feces.

Diagnostic problems. — Sometimes confused with the eggs of *H. nana*, the mouse tapeworm, which also may infect humans. The eggs of *H. nana* are smaller (30–47 μ in diameter) and they contain polar filaments arising from the membrane around the oncosphere.

FIGURES. — Plate 70:1–4; also see Plate 69:4.

PLATE 70

Hymenolepis diminuta, Eggs

Figs 1 and 2. As illustrated in these two figures, the *H. diminuta* egg is spherical and clearly different from that of *H. nana*. It is larger, has a thicker shell, is bile-stained, and the inner envelope surrounding the oncosphere lacks polar filaments.

Fig 3. In this low-power field, there are two eggs exhibiting marked differences in coloration of the outer shell. In these instances, lightly colored eggs should not be confused with those of *H. nana*.

Fig 4. *H. diminuta* and *H. nana* eggs are seen in the same field. The morphologic differences between the two species are clearly evident.

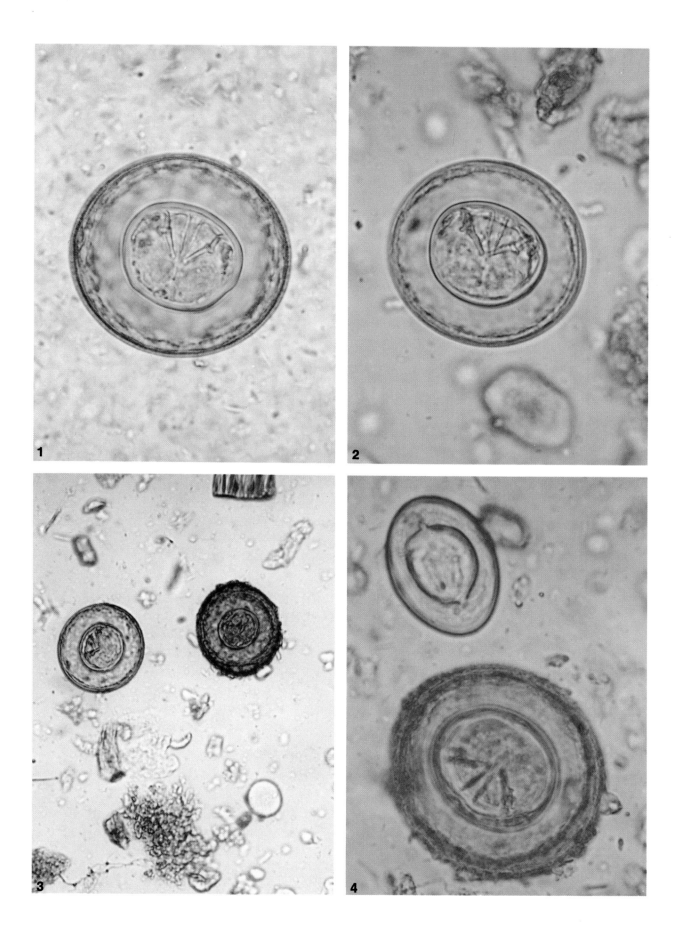

Dipylidium caninum

CLASSIFICATION. — Cestode.

DISEASE. — Dog tapeworm, double-pored tapeworm infection.

GEOGRAPHIC DISTRIBUTION. — Cosmopolitan.

LOCATION IN HOST. — Small intestine.

MORPHOLOGY. — Adult worms. — Adult tapeworms are 10–70 cm long. The scolex is conical, has four suckers, and has a retractile rostellum armed with several rows of small spines. Gravid proglottids are longer than wide (23 × 8 mm), and are divided into compartments, each containing 8–15 six-hooked eggs. There are two genital pores, one at each lateral margin.

Eggs. — These thin-shelled eggs measure 25–40 μ in diameter and they contain six hooks.

LIFE CYCLE. — Gravid proglottids usually are shed in feces or emerge spontaneously from the anal opening. Flea larvae ingest these proglottids, and the six-hooked embryos develop into cysticercoids in the body cavity of the flea larvae first and remain there as the adult flea matures. Infection is acquired by ingestion of infected fleas. Maturation to the adult tapeworm takes three to four weeks. Dogs are the usual host for this parasite but human infections, especially those in children, probably are more frequent than recorded in the literature.

DIAGNOSIS. — Usually by demonstration of the characteristic double-pored gravid proglottids in feces or after spontaneous passage. Occasionally proglottids will rupture and packets of eggs with the characteristic six hooks will be present in the feces.

COMMENTS. — Infection usually occurs in humans who have close association with infected dogs and their fleas.

FIGURES. — Plate 71:1–4.

PLATE 71

Dipylidium caninum, Adult Worms and Eggs

Fig 1. The scolex of this carmine-stained, adult cestode has four large suckers and a conical rostellum bearing many rows of small spines.

Fig 2. This gravid, carmine-stained proglottid illustrates the compartmentalized nature of the uterus and the two, laterally placed, genital pores so characteristic of the species. *D. caninum* is a common parasite of dogs, and chains of gravid proglottids may be passed. In their natural state, these proglottids may resemble grains of rice.

Figs 3 and 4. Egg packets in fecal smears. The typical egg packet of *D. caninum* is shown in Figure 3. These packets usually contain 8–15 eggs, each of which contains an oncosphere. The oncospheres, each with their six hooks, are seen best at higher magnification in Figure 4.

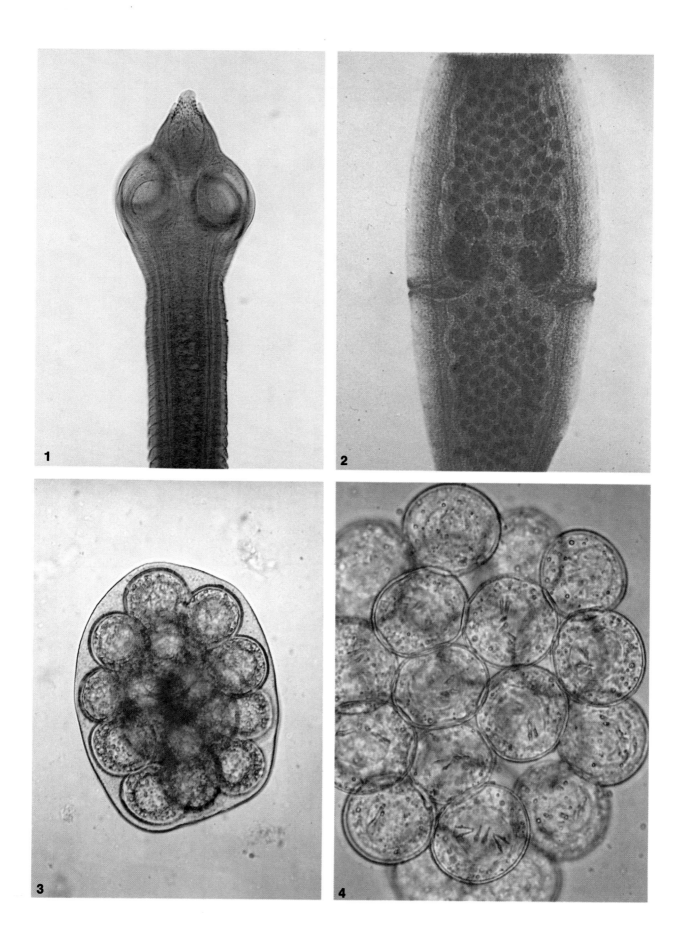

Echinococcus granulosus

CLASSIFICATION. — Cestode.

DISEASE. — Hydatidosis, hydatid disease.

GEOGRAPHIC DISTRIBUTION. — In sheep and cattle-raising countries of the world — especially Australia, New Zealand, southern Africa, southern South America, parts of Europe, with sporadic foci in North America, and the Orient.

LOCATION IN HOST. — Only the larval stage infects humans. Hydatid cysts may form in any organ or tissue including bone, but are most common in the liver, lung, and central nervous system. Adult tapeworms occur in the small intestine of dogs and other canids.

MORPHOLOGY. — Adult worms. — Adult tapeworms are very small, 3–6 mm long, and usually consist of a scolex and three proglottids. The scolex has four suckers and a rostellum with hooklets. The single, gravid proglottid is longer than it is wide, and is filled with typical, *Taenia*-like eggs.

Eggs. — Morphologically identical to *Taenia* eggs: spherical, 31–43 μ in diameter, and yellow-brown in color. The shell is thick and radially striated, giving a prismatic appearance. A thin, outer, primary membrane occasionally can be seen around some eggs. The egg contains a six-hooked embryo called an *oncosphere*.

Larval stage. — In human tissues, hydatid cysts consist of an outer wall differentiated into an outer laminated, nonnucleate layer and an inner layer of germinal epithelium. From the germinal layer, masses of cells bud into the cystic cavity and form scoleces and brood capsules. Scoleces invaginate into their own bodies and brood capsules have numerous scoleces budding from their walls into their own smaller cystic cavity. Death and degeneration of scoleces and brood capsules results in fluid accumulation and disintegrating cestode tissue, in-cluding hooklets, in the cyst. This material is referred to as *hydatid sand*. Hydatid cysts may vary considerably in size depending on the age of the cyst and the anatomic location in which they are situated.

LIFE CYCLE. — Dogs and other canids are the definitive hosts, and they excrete eggs in feces. When these eggs are ingested by cattle, sheep, or other animals, the oncosphere migrates from the intestinal lumen to various parts of the body (especially to the liver and lungs) and develops into the hydatid cyst. If the cyst is ingested by canids, the numerous scoleces in the cyst will each become a minute adult in the intestine. The human is essentially a dead-end host for the hydatid cyst.

DIAGNOSIS. — Human infection is diagnosed mainly on clinical grounds, including a history of possible exposure to the infection. Hydatid cysts can be demonstrated by radiology, laparotomy, or accidentally during surgical procedures. Serologic tests are useful in diagnosis. Infections in dogs sometimes are difficult to diagnose, since there are several species of taeniid tapeworms that can parasitize these animals, and their eggs are identical to those of *E. granulosus*.

Diagnostic problems. — As just outlined, human infections sometimes are difficult to diagnose. Aspiration of cysts to demonstrate characteristic hydatid sand is a dangerous procedure that should be performed only under exceptional circumstances.

COMMENT. — In the western part of the United States, increasing numbers of *E. granulosus* cases in humans have been reported. In addition to *E. granulosus*, there are other species of the genus *Echinococcus* infecting canines and felines that may cause various types of hydatid cyst formation in humans.

FIGURES. — Plate 72:1–4.

PLATE 72

Echinococcus granulosus, Adult Worms and Hydatid Sand

Fig 1. Adult worm, carmine stain. This tapeworm of canids is unusual in that it is extremely small (3–6 mm long) and usually is composed of a scolex with a crown of hooks and three or more proglottids: one immature, one mature, and one or two gravid. Although the adult worm never matures in humans, the larval stage infects humans and causes hydatid disease. The eggs of this species are indistinguishable from those of the genus *Taenia* and they are infective to humans.

Figs 2 and 3. Hydatid sand. Aspiration of fluid from a hydatid cyst may yield numerous scoleces that bear hooklets and are referred to as *hydatid sand*. Characteristically, the scoleces are invaginated into their own bodies (Fig 2) or, if the material is placed into saline, the scoleces may evaginate (Fig 3).

Fig 4. Hooklets. The characteristic morphology of the rostellar hooks is best seen in a squash preparation of the scolex. In aspirated hydatid fluid, the individual hooks may be found and can be recognized by their morphologic features.

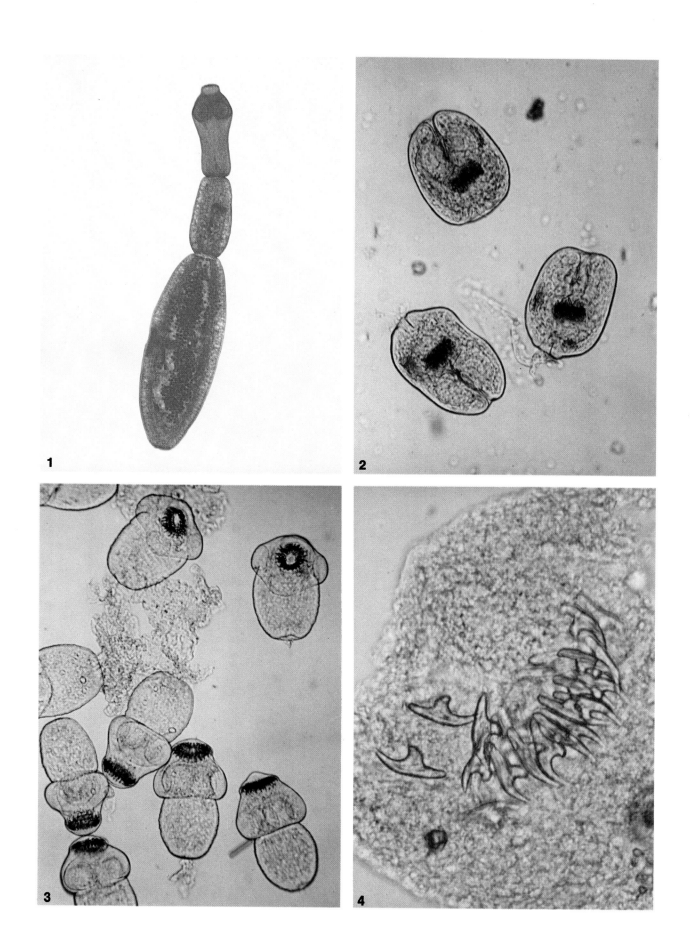

Uncommon Cestode Parasites in Humans

Two examples of animal tapeworm infections that sometimes occur in humans are species of the genera *Bertiella* and *Mesocestoides*.

Species of *Bertiella* occur in monkeys and are transmitted by mites. Their eggs are approximately 38–46 μ by 33–43 μ and are characterized by their pyriform apparatus—an inner chitinous membrane that has a pair of filamentous projections around the six-hooked embryo.

Mesocestoides species are parasites of carnivores and their larval stages occur in the tissues of reptiles, birds, and small mammals. *Mesocestoides* eggs have membranous walls surrounding the six-hooked embryo and they are quite small, measuring only 19–24 μ in length.

FIGURES.—Plate 73:1–4.

PLATE 73

Uncommon Cestode Parasites in Humans

Figs 1 and 2. *Bertiella* eggs, unlike other human tapeworm eggs, have a pyriform apparatus surrounding the six-hooked oncosphere. In spite of the unique morphology of this egg, the six hooks readily identify it as a tapeworm parasite.

Figs 3 and 4. These small *Mesocestoides* eggs have a thin limiting membrane surrounding the six-hooked oncosphere. They frequently may be seen in the feces of dogs.

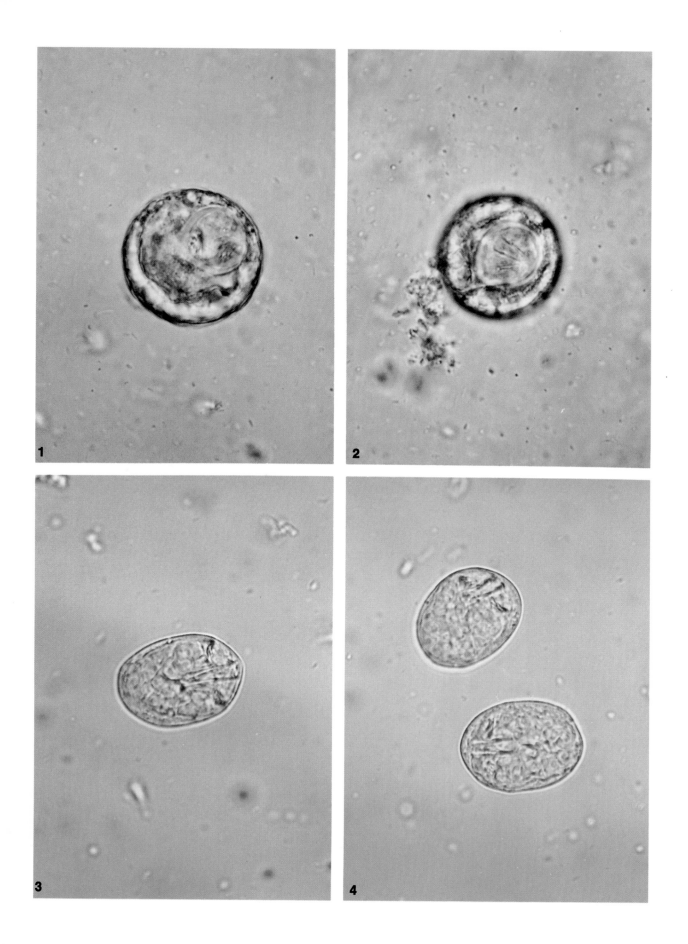

Recommended References

TEXTS — MEDICAL PARASITOLOGY AND TROPICAL MEDICINE

Binford CH, Connor DH: Pathology of Tropical and Extraordinary Diseases. An Atlas, vols 1 and 2. Washington DC: Armed Forces Institute of Pathology, 1976

Faust EC, Beaver PC, Jung RC: Animal Agents and Vectors of Human Disease, ed 4. Philadelphia: Lea & Febiger, 1975

Faust EC, Russell PF, Jung RC: Craig and Faust's Clinical Parasitology, ed 8. Philadelphia: Lea & Febiger, 1970

Hunter GW, Swartzwelder JC, Clyde DF: Tropical Medicine, ed 5. Philadelphia: WB Saunders Co, 1976

Marcial-Rojas RA (ed): Pathology of Protozoal and Helminthic Diseases with Clinical Correlation. Baltimore: Williams & Wilkins Co, 1971

Markell EK, Voge M: Medical Parasitology, ed 4. Philadelphia: WB Saunders Co, 1976

Muller R: Worms and Disease. A Manual of Medical Helminthology. London: William Heinemann Medical Books Ltd, 1975

Wilcocks C, Manson-Bahr PEC: Manson's Tropical Diseases, ed 17. Baltimore: Williams & Wilkins Co, 1972

MANUALS AND TEACHING AIDS

Garcia LS, Ash LR: Diagnostic Parasitology. Clinical Laboratory Manual, ed 2. St Louis: CV Mosby Co, 1979

Melvin DM, Brooke MM: Laboratory Procedures for the Diagnosis of Intestinal Parasites. USDHEW publication no. (CDC) 75-8282, 1974 (NTIS, US Dept of Commerce, Springfield VA, no. PB-297 958/PTX)

Smith JW, et al: Blood and Tissue Parasites. Diagnostic Medical Parasitology, vol 1. Chicago: American Society of Clinical Pathologists, 1976a

Smith JW, et al: Intestinal Protozoa. Diagnostic Medical Parasitology, vol 2. Chicago: American Society of Clinical Pathologists, 1976b

Smith JW, et al: Intestinal Helminths. Diagnostic Medical Parasitology, vol 3. Chicago: American Society of Clinical Pathologists, 1976c

Spencer FM, Monroe LS: The Color Atlas of Intestinal Parasites, rev ed. Springfield IL: Charles C Thomas Publisher, 1975

Index